Microstructure of Smectite Clays and Engineering Performance

T0203816

Microstructure of Smectite Clays and Engineering Performance

Roland Pusch and
Raymond N. Yong

CRC Press
Taylor & Francis Group
Boca Raton London New York

CRC Press is an imprint of the
Taylor & Francis Group, an **informa** business
A TAYLOR & FRANCIS BOOK

CRC Press
Taylor & Francis Group
6000 Broken Sound Parkway NW, Suite 300
Boca Raton, FL 33487-2742

First issued in paperback 2019

© 2006 Roland Pusch & Raymond N. Yong
CRC Press is an imprint of Taylor & Francis Group, an Informa business

No claim to original U.S. Government works

Typeset in Sabon by
Newgen Imaging Systems (P) Ltd, Chennai, India

ISBN13: 978-0-415-36863-6 (hbk)
ISBN13: 978-0-367-44641-3 (pbk)

British Library Cataloguing in Publication Data
A catalogue record for this book is available
from the British Library

Library of Congress Cataloging in Publication Data
A catalog record for this book has been requested

Visit the Taylor & Francis Web site at
http://www.taylorandfrancis.com

and the CRC Press Web site at
http://www.crcpress.com

Contents

Figures

Figures

Colour plates

The following colour plates appear at the end of the book.

Colour plates

The following colour plates appear at the end of the book.

Tables

Preface

Studies on the properties and behaviour of clay soils, conducted in Soil Engineering, have revealed the importance of Soil Structure in the control of the properties and response performance of clays. Early descriptions of clay soil structure, in the 1950s–1960s, using such terms as 'honeycomb', 'cardhouse', 'flocculated', 'dispersed' and 'oriented' testify to the level of descriptive scrutiny of the micromorphology of clay soils. Many of these descriptive terms were based on evidence or deductions gained using thin sections for light microscopy and controlled experiments on various kinds of clays. With more sophisticated techniques presently available, for example transmission electron microscopy (TEM) and scanning electron microscopy (SEM) and nuclear magnetic resonance, and with considerable input from colleagues in Soil Science and Clay Mineralogy working on the properties and characteristics of clays, we have learnt more not only about soil structure but also about the importance and role of the microstructural units (MUs) that make up the total soil structure. The arrangement of these MUs, (domains, peds, aggregate groups, etc.) and the bondings established between them are key to the macrostructural integrity of a soil.

The growing use of swelling clays, and particularly smectite clays, as integral components in engineered clay barriers for waste isolation, has fuelled the need for a body of literature dealing with the properties of smectites and their performance under waste isolation scenarios. The particular situation of isolation of canisters containing high level radioactive waste (HLW) presents us with some very challenging issues with respect to the use of the smectites as buffer and backfill material. To meet the challenges, we need to understand how the clay material will respond to chemical and heat attack, and to other degradative forces associated with abiotic and biotic reactions and particularly under the a very long time span (hundreds of thousands of years) and high temperature. Our individually separate and collective work with swelling clays over the past 40 plus years have shown the importance of developing an understanding of the layer-lattice structure of the basic clay mineral, and particularly on how the mineral responds to the various external forces and agents. The properties and characteristics of the mineral that give it its swelling behaviour are those very same properties and characteristics that are seen to be highly dependent on the microstructural constitution of the soil.

Since the primary focus of the book is with respect to the use of smectite clays as buffer and barrier materials for waste isolation in general, and for isolation of HLW in particular, no attempt is made to provide a detailed background on the mineralogy and structure of smectites. Instead, a general overview of the basic elements constituting the structure of phyllosilicates is given so as to lay the basis for the performance of the layer-lattice material upon water uptake (Chapter 2). The structure and performance of smectite clays in particular, and clays in general, are considered in Chapter 3. Since direct examination of MUs is not possible when these soils are undergoing stress or are being subject to external inputs, we need to establish deductive techniques that would give us insight into microstructural performance using bulk macroscopic data (Chapter 5). The aim of these chapters is to provide a microstructural basis for evaluating performance of the material. Since the principal role of the compacted smectite clay in a barrier system is to 'manage' the flow of fluids and contaminants, the available flow paths and the processes restricting advective and diffusive flow are important factors to be considered. Microstructrural control on these becomes important (Chapter 6). Chapter 4 uses the background information provided in the previous chapters and gives guidance on how these can be maximized.

We address the use of these clays as buffer and barrier systems for waste isolation, in general and in particular for the HLW isolation problem, using a specific underground HLW repository as an example (Chapter 7). Since we are expected to design and construct isolation barriers for HLW isolation that would function effectively for hundreds of thousands of years, the use of analytical-computer models to predict the durability and capability of such barriers is discussed in Chapter 8. The example in Chapter 7 is used as the focus of the modelling efforts expended by various representative groups involved in the isolation of HLW.

We would like to express our gratitude to our colleagues working in the fields of Soil Engineering and Soil Science, and to the Swedish companies Swedish Nuclear Fuel and Waste Management Co, Stockholm, and Ragn-Sell Waste Management AB, Hoegbytorp. We are grateful to them for their inputs in our valuable discussions with them, for their encouragement, and also for allowing the use of data emanating from projects conducted by them. We extend our gratitude also to the European Commission (EC), which financed, in part, international research projects such as the Prototype Repository Project and the Cluster Repository Project (CROP), which provided us with rich data sources.

Roland Pusch
Lund Sweden

Raymond N. Yong
North Saanich, B.C. Canada
February, 2005

1 Introduction

Soil mechanics investigators dealing with classical geotechnical problems such as slope stability and settlement of buildings, have always sought to understand the various mechanisms and physico-chemical processes upon which to develop their mathematical models for prediction of soil behaviour. This led to the fruitful cooperation between mineralogists and soil engineers exemplified by I.Th. Rosenqvist and L. Bjerrum in Norway, and by T.W. Lambe and various soil physicists such as H. van Olphen and P.F. Low in the United States half a century ago. Their work deepened our understanding of the relationships between the microstructure and bulk physical properties of illitic and kaolinitic clays, especially in respect to the importance of the physical state of the porewater. Much less was known about smectitic clays – to a very large extent because of the limited use of the material. However, in the past two decades emphasis on the use of swelling clays for buffer and barrier systems in the containment and isolation of hazardous and non-hazardous waste materials requires us to obtain a better understanding of such clays. A considerable portion of the material contained in this book has been developed in conjunction with international and Swedish research and development work on the requirements for safe isolation of high level radioactive waste (HLW).

A group of swelling clays that well suits the buffer–barrier requirements for containment of landfills and other facilities needing liner systems is the smectite family. Such being the case, and especially in the context of isolation of radioactive waste exposed to a special and demanding environment of high temperature and radiation, the need for a better understanding of the behaviour of smectite clays is paramount. There exists a critical requirement for knowledge and assurance of competent performance of the smectite clays over the long time period (tens of thousands of years) of safe and secure isolation of the HLW. The long time period for performance assurance is a significantly more important parameter than in conventional soil mechanics. This has led to some very comprehensive research studies on the microstructure of the smectitic clays, and its importance in the physical behaviour of such clays – especially in respect to the influence of chemical factors. The outcome of past and ongoing studies can be seen in the many research reports issued by the international organizations charged with the responsibility for the safe disposal of radioactive waste in Europe, such as the Swedish Nuclear Fuel and Waste Management Co (SKB).

One such example is the European Commission (EC) study 'Microstructural and chemical parameters of bentonite as determinants of waste isolation efficiency' (Contract F14W-CT95-0012). This study has shown that the microstructural properties and characteristics (features) are predominant factors in the determination of the isolation potential of bentonite-based buffers and backfills.

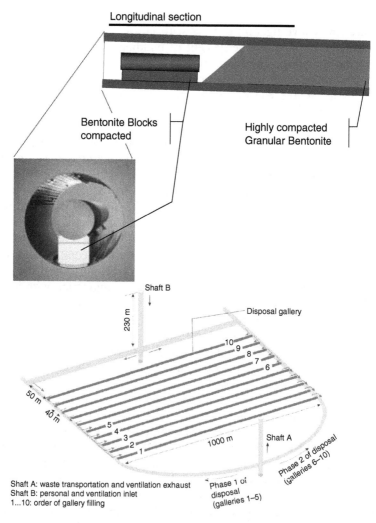

Figure 1.1 Example of concept for disposal of HLW in argillaceous rock (NAGRA, Switzerland).

Theoretical modelling of the performance of smectite clays as buffers and backfills in underground repositories designed to isolate HLW is presently being seriously considered in many international research projects. These will be discussed in some detail in this book. However, the necessary foundation for the models, that is, reliable conceptual modelling, has yet to be properly laid. The real field problem and its various processes have yet to be fully and correctly posed – as will be obvious to the reader of this present book.

Much of the work that is being undertaken on the role of microstructure on the physical properties of smectite clays for waste isolation can be seen to have a considerable impact on such applications as low-temperature soil mechanics, storage of gas and road design and maintenance. To elaborate on the technical aspects of the role of microstructure in smectite soil behaviour, some of the results from ongoing research will be briefly discussed. We use the term *smectite clay* to mean a clay soil containing only or primarily smectite clay minerals, and the term *smectitic clay* to mean a clay that has

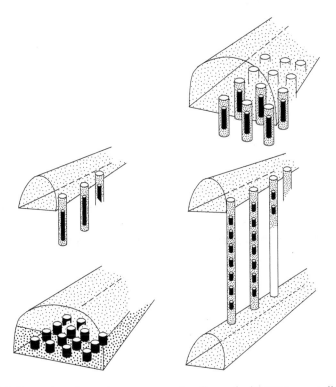

Figure 1.2 Examples of repository concepts for disposal of HLW in crystalline rock. Upper right: Multiple deposition holes in large vault in crystalline rock and salt. Mid left: SKB's KBS-3V concept. Lower left: Containers embedded in large vault. Lower right: Several canisters in medium-long vertical or subhorizontal holes from drifts [1].

Figure 1.3 Large block of highly compacted bentonite powder being placed in bored
        deposition hole with 1.75 m diameter and 8 m depth for surrounding a
        HLW canister.

the smectite clay mineral amongst the other clay minerals that constitute the
clay material. Such a clay material would be a swelling clay, but its swelling
behaviour would be less pronounced in comparison to a smectite clay of the
same porewater chemistry.

The two driving forces in respect to the containment and isolation of
HLW are safety and economy. The response to the driving forces has been
to comprehensively investigate the use of smectitic clays as part of the
containment and isolation system. This is illustrated by the extensive
research study programmes in many different countries that have led to
concepts such as those shown in Figures 1.1–1.3. These kinds of studies
have led to a greater understanding of the performance of smectitic clays at
the microstructural level – forming the basis for conceptual and theoretical
models that are needed for the design of clay-based engineered barriers.

1.1   Reference

1. Pusch R., 1994. *Waste disposal in rock. Developments in geotechnical engineering, 76.* Elsevier Scientific Publishing Company, Amsterdam–Oxford–New York.

2 Smectite clays

2.1 Particles

2.1.1 *Crystal constitution*

The minerals that give smectite clay its well-known valuable waste–isolation properties of low-permeability and expandability consist of stacked lamellae each of them consisting of two sheets of SiO_4 tetrahedrons confining a central octahedral layer of hydroxyls and Fe, Mg or Li ions. These minerals (Figures 2.1 and 2.2) belong to the group known as:

- *Montmorillonite* – Only Si in the tetrahedrons and Al in the octahedrons.
- *Beidellite* – Si and Al in the tetrahedrons and Al in the octahedrons.
- *Nontronite* – Si and Al in the tetrahedrons and Fe in the octahedrons.
- *Hectorite* – Si in the tetrahedrons and Mg and Li in the octahedrons.
- *Saponite* – Si and Al in the tetrahedrons and Mg in the octahedrons.

The simplified views of the tetrahedron and octahedron are commonly used to represent the elements of a basic unit cell (middle schematic in Figure 2.2). The literature uses both lamellar and layers interchangeably. Thus we will often see such notations as *layers* to mean *lamellae*, and *inter-layer* space to mean *interlamellar* space – as shown in Figure 2.2. A typical feature of these minerals is that the lamellae usually do not exist as single units but combine to form stacks (as shown in the bottom portion of Figure 2.2) with 3–10 aligned coupled lamellae – depending on the bond strength, which is different for different adsorbed cations. Such stacks represent *clay particles*, which are more or less interwoven in undisturbed bentonites. This interwoven feature not only makes it difficult to define the granulometric properties of such clays, but also results in the production of different particle size distributions depending on the manner of sample procurement and testing.

It is generally believed that there are vacations and replacement of the ions in the ideal smectite types by ones that have different valences. The net result of these is the development of a net negative charge associated with

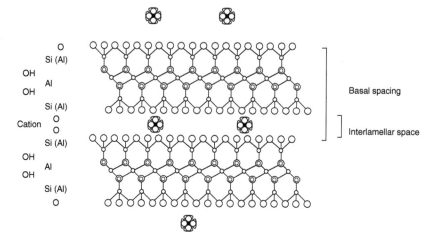

Figure 2.1 Commonly assumed crystal constitution of a smectite lamella, which is about 1 nm (10 Å) thick in dehydrated state. The space between the two tetrahedral sheets is commonly called the interlamellar ('internal') space while the space between stacks of lamellae is termed 'external'.

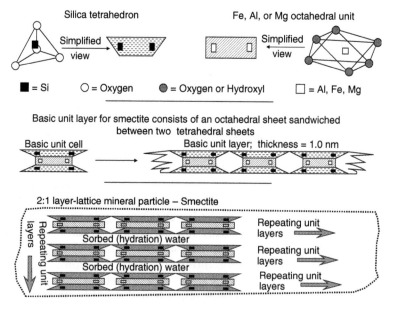

Figure 2.2 Top schematic shows a simplified representation of silica tetrahedral and octahedral unit. Middle section shows the basic unit cell and the basic unit layer with thickness of about 1 nm (10 Å). Bottom section shows the stacking of unit layers with the interlayer or interlamellar space occupied with hydration water. Note that the terms *layer (layers)* and *lamella (lamellae)* mean the same thing and are often used interchangeably.

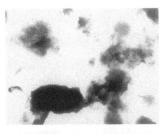

1 μm

Figure 2.3 Micrographs of smectite (montmorillonite) particles. Upper: TEM of aggregated crystallites with a maximum size of about 0.5 m. An illite particle with about 1 μm length is seen in the lower left of the graph (dark). Lower: SEM. Notice the difficulty in dispersing smectite into individual lamellae.

the particles, and is the reason for adsorption of cations that can be more or less strongly hydrated. However, as we will touch on later, alternative crystal concepts exist that imply that proton exchange could be responsible for the cation-exchange properties of the mineral [1].

2.1.2 Morphology

Smectite particles have a peculiar shape and appear in networks that make it difficult to identify individual members and define their size and shape (Figure 2.3). Effective dispersion breaks up the networks and causes partial separation of the lamellae. Determination of the size distribution of a smectite clay is therefore largely a matter of how the sample is prepared.

2.2 Basic properties of smectite particles

2.2.1 Interlamellar hydration of smectite particles

The hydration of the interlamellar space depends on the charge and coordination of adsorbed cations and water molecules [2]. A water hull, one monolayer thick and containing cations, is sorbed on the basal surfaces

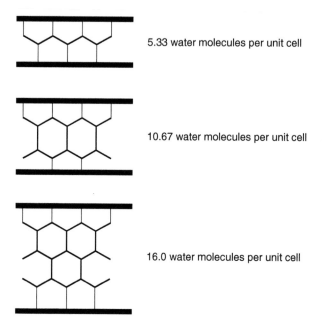

5.33 water molecules per unit cell

10.67 water molecules per unit cell

16.0 water molecules per unit cell

Figure 2.4 Model of hydrate configuration in interlamellar montmorillonite with Edelmann/Favejee crystal constitution. The stable monolayer yields a d-spacing of 17.81 Å and a water content w = 11.9%, while for two layers d = 22.41 Å and w = 23.8%. For three layers d = 25.17 Å and w = 35.7% [3].

of practically all minerals, and plays a role in diffusive ion transport and in osmotic swelling of smectites. The interlamellar hydrates largely determine the swelling potential, whereas the surface properties of the stacks of lamellae almost entirely determine their plasticity and rheological behaviour. Figure 2.4 shows a schematic model of the organization of water molecules in the interlamellar space of two theoretically possible crystal constitutions of montmorillonite with Li or Na in interlamellar positions [3]. Figure 2.5 shows a proposed organization of water and Ca ion complexes [4, 5].

The coordination of interlamellar cations, the crystal lattice and the water molecules depend strongly on the size and charge of the cations and of the charge distribution in the lattice. Three hydrate layers can only be formed in some of the smectite species, and normally only when sodium, lithium or magnesium are in interlamellar positions (Table 2.1). One finds from this table that montmorillonite is the only smectite mineral of the three mentioned that can expand by forming three interlamellar hydrates.

In water, hydration proceeds until the maximum number of interlamellar hydrates are formed, provided that there is no geometrical restraint. In humid air, the number of hydrates obtained depends on the relative humidity (RH) (Table 2.2). This latter table shows that one interlamellar hydrate

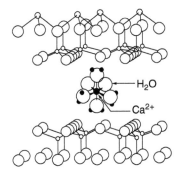

Figure 2.5 Interlamellar hydrated Ca ion complex with water molecules strongly attached to the cation and interacting also with the siloxane oxygens. It forms the well-known two-layer hydrate in Ca-montmorillonite [4].

Table 2.1 Number and thickness of interlamellar hydrates in Å [2]

Smectite clay	M[a]	1st hydrate	2nd hydrate	3rd hydrate
Montmorillonite	Mg	3.00	3.03	3.05
	Ca	3.89	2.75	—
	Na	3.03	3.23	3.48
	K	2.42	3.73	—
Beidellite	Mg	2.69	2.69	—
	Ca	2.30	2.30	—
	Na	2.15	2.15	—
	K	2.54	—	—
Nontronite	Mg	2.92	3.00	—
	Ca	3.05	3.37	—
	Na	2.70	2.79	—
	K	2.60	—	—

Note
a Interlamellar adsorbed cation.

is formed in Na montmorillonite when the RH is approximately in the interval 20–40%, while two hydrates require 40 < RH < 60%. For RH exceeding about 60–70% three interlamellar hydrates are formed.

Table 2.1 shows that beidellite and nontronite can form no more than two interlamellar hydrate layers irrespective of the type of adsorbed cation, and it therefore has a lower swelling potential than montmorillonite with sodium or magnesium in the interlamellar exchange positions. With potassium or calcium in these positions, a third hydrate does not form in this or any other smectite mineral because of the charge and size of these ions. K^+ is deeply and strongly anchored in the cavities of the siloxane (Si–O) lamella because it fits geometrically and stereochemically [4].

Table 2.2 d-spacing and water absorption of Na-montmorillonite in atmosphere with different RH

RH %	Adsorption area, m²/g solids	d(001) spacing, Å	Water adsorption in g/g solids, %. Pure clay	Water adsorption in g/g solids, %. MX-80
14	33	10	4.5	3.2
19	33	10	6.5	4.6
24	440	12.4	8.0	5.6
38	440	12.6	11.5	8.1
60	810	14.2	17.0	12.0
70	810	15.0	21.5	15.0
80	810	15.2	24.0	17.0
90	810	15.4	28.0	20.0
95	810	15.4	36.5	26.0

The table 2.2 includes a column for commercial Na-dominated bentonite (MX-80), which contains about 70% montmorillonite, the rest being rock-forming minerals. One finds that the water adsorption logically is about 70% of that of pure Na-montmorillonite.

When Na is in the interlamellar space the coupling to water molecules is probably weak and the cations relatively free to move. According to some researchers, the interlamellar water molecules form their own H^+-bonded structure (Figure 2.4), while in Ca-smectite the cations are strongly hydrated and the interlamellar complexes rigid and stable (Figure 2.5). This means that the swelling pressure and the pressure required to expel inter-lamellar water is higher for Ca-smectite than for Na-smectite for bulk densities exceeding about 2000 kg/m³. The much higher mobility of Na ions is verified by the significantly higher diffusion coefficient – as is discussed later in this chapter.

The nature of the water structure in the interlamellar space is still a matter of dispute. However, there are reasons to believe that at least for Li and Na in the interlamellar space, the crystal structure may be different from that in Figure 2.1, implying a certain number of apical tetrahedrons – as indicated in Figure 2.6. This is supported by nuclear magnetic resonance ('magic angle spin-echo') studies that show an obvious difference in the co-ordination of the silicons in Na- and Ca-montmorillonite. These indicate that the first-mentioned is of the type in Figures 2.1 and 2.4 and the latter as in Figure 2.6 [6].

From a practical point of view, it is believed that the physical nature of the interlamellar water is different because the stacks of lamellae in Ca-montmorillonite (i.e. the clay 'particles') are much stronger and less easily disrupted than in Na-montmorillonite. This provides the first-mentioned with a character similar to that of a silty, frictional soil with less potential to form gels. A unit cell of Ca-montmorillonite is expected to be stiffer with less creep strain than one of Na-montmorillonite. In both types, the

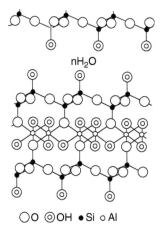

nH₂O

○○ O ◎○OH ●Si ○Al

Figure 2.6 Possible crystal structure for Li- and Na-montmorillonite with apical tetrahedrons (Edelmann/Favejee/Forslind) [6].

interlamellar water is believed to be more viscous than ordinary water and to be largely immobile under normal hydraulic gradients.

The interlamellar space offers large amounts of hydration sites due to the crystal lattice constitution. This determines the charge and coordination of adsorbed cations and water molecules. The hydration properties control the swelling potential, plasticity and rheological behaviour, and are hence of fundamental importance.

The coordination of interlamellar cations, the crystal lattice and the water molecules are strongly dependent on the size and charge of the cations, and on the charge distribution in the lattice – as discussed earlier. When submerged in free water, the hydration of an initially dry montmorillonite crystallite proceeds until the maximum number of hydrates is formed provided that there is no geometrical restraint. In humid air, the number of hydrates depends on the relative humidity since a balance exists between the water molecule concentrations in the moist air and on the basal surfaces of the crystallites as well as between the hydration stages of the basal surfaces and the interlamellar space.

Spin–echo proton measurements allow us to distinguish between intra- and extralamellar water. These measurements suggest that the relaxation time T_2 is around 20–40 µs for protons in interlamellar water in montmorillonite as compared to 2.3 seconds in free water [6]. This obvious difference indicates strong structuring and very limited mobility of interlamellar water. The small difference in proton mobility of one and three hydrate layers indicates that the interlamellar water has approximately the same physical properties irrespective of the number of hydrates.

2.2.2   Extralamellar hydration – the basal planes of smectite stacks

External surfaces are the basal planes and edges of the stacks of lamellae. The first-mentioned consists of hexagonal arrangements of oxygens or hydroxyls, and can attach water molecules by establishing hydrogen bonds (Figure 2.7). When basal planes come close to each other, interacting diffuse electrical double-layers are formed with a relatively high concentration of cations near the surface. This affects the organization and physical state of the hydrates (Figure 2.8). Thus, it is possible that the first few hydrate layers on smectite surfaces (Water A) are more viscous than those in Water B, which are in turn less viscous than normally structured Water C. These issues are of fundamental importance in the understanding of diffusion processes in smectite clays.

2.2.3   Hydration at elevated temperature

The influence of temperature and water pressure on water sorption and particle thickness of montmorillonite clay confined in autoclaves under controlled water pressure conditions have been examined by several investigators. Colten [8] reports on a typical, carefully conducted test consisting

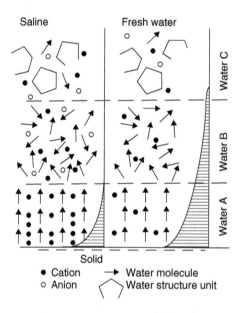

Figure 2.7 Proposed organization of water and ions at clay mineral surfaces. Water A is largely immobile through coupling to surface atoms. Water B is low-viscous water. Water C is free with normal viscosity. Hatched areas represent the charge distribution in the electrical double-layers (left: marine, right: fresh-water) [6].

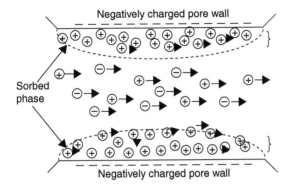

Figure 2.8 Interacting electrical double-layers [7].

of a series of experiments on montmorillonite slurry with a fluid content of 800% (density 1050 kg/m³) placed in a cell exposed to temperatures up to 200°C and to hydraulic pressures of up to 40 MPa. The molality of the NaCl solution that was used for saturation was 1–5. This study showed that the change in interlamellar spacing was negligible for all the different tests. The reasoning given for the negligible change was that there was no effective pressure persisting in the clay or on its boundaries.

In the Pusch–Karnland [5] series of experiments on MX-80 clay, the clay was saturated with distilled water and enclosed in autoclave cells. This resulted in development of high porewater overpressures (i.e. high excess porewater pressures) because of the heating process. The overpressure was 40–70 MPa for temperatures up to 200°C and the swelling pressure of the clay samples, which had a density of 1300 kg/m³, was about 70 kPa. X-ray diffraction (XRD) analysis after extraction of the clay and preparation by sedimentation in the usual fashion showed very small changes in inter-lamellar spacing, that is, from 16.7 Å for room temperature reference samples to 16.7–17.3 Å spacings (001) for samples heated for 0.5 years at 150°C and 200°C, respectively.

The Pusch–Karnland study confirmed that no permanent contraction or compression takes place under very low effective pressures in the course of heat treatment. However, the peak heights (in XRD) were increased by 5% and 50%, respectively, indicating that alignment of stacks of lamellae had in fact taken place. This process was probably caused by compression and reorientation of the stacks under the effective pressure that actually existed. It is believed that compression of the stacks of lamellae was triggered by heat-induced reduction of the strength of the interlamellar hydrate layers. Permanent reduction in interlamellar spacing can result from cementation of precipitated compounds but the initial spacing will otherwise be restored.

2.2.4   *The chemical potential*

The influence of chemical potential of water μ_0 is of fundamental importance for the build-up of porewater pressure and performance clay soils. Water with few dissolved ions has a higher chemical potential than water with solutes. The drop in potential for increasing the ion concentration can be expressed as in Eq. (2.1) [9]:

$$\mu_0 = -kTCv \qquad (2.1)$$

where k is the Boltzmann's constant, T is the temperature, C is the cation concentration in hydrates formed at mineral surfaces and v is the partial volume of water.

The cations cause polarization of neighbouring water molecules resulting in decreased mobility of the molecules and a reduction of the chemical potential. Further reduction is due to van der Waals attraction between the hydrates and adjacent mineral lattice, and to hydrogen bonds established between the hydrates and lattice hydroxyls. Interlamellar hydrates are pressed together by the attraction between the lamellae. This increases the chemical potential. It is therefore highest for low water contents and higher for interlamellar hydrates than for hydrates on basal surfaces of clay particles. Nakano *et al.* [9] found that the chemical potential of the first hydrate adsorbed on basal surfaces is on the order of E7 J/kg, and about E2 J/kg for hydrates at around 60 Å from the surfaces.

This theory can be used for calculating water retention curves in the form of relationships between the water content and the chemical potential (Figure 2.9). The values naturally depend on the number of lamellae. This is illustrated by the much higher chemical potential (3E4 J/kg) for the volumetric water content 20% when the number of lamellae is 8 than when this number is 4 (5E3 J/kg). The water retention of dense clay is hence

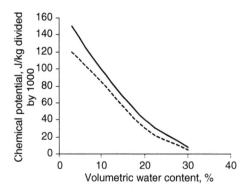

Figure 2.9 Chemical potential of montmorillonite-rich clay calculated by use of molecular dynamics and determined experimentally [9]. Upper: calculated, lower: experimental.

higher than that of soft clays. Similarly, the chemical potential of the porewater is much higher for dense than for soft clay. For a water content of about 20% the chemical potential 10^5 J/kg corresponds to a suction of as much as 100 MPa.

Molecular dynamic/homogenization theories have been used to model the organization of montmorillonite lamellae and interlamellar water. These have led to some interesting microstructural features [10]. The decrease in the chemical potential of the porewater due to the solute, electrical field and van der Waals forces implies that for a bulk dry density of 1000 kg/m³, the stacks contain 4 lamellae whereas for a dry density of 1800 kg/m³ this number is 8. The water adjacent to montmorillonite surfaces is concluded to be rigidly associated with the crystal lattice. Further (additional) hydrates will have viscosities that drop quickly with increasing distance from the surfaces – up to 25 Å. Beyond this distance, the viscosity is the same as for free water (Figure 2.10).

A study [3] using vapour pressure measurements and XRD technique supports the conclusion that three distinct hydrate layers are established in the interlamellar space in montmorillonite. It further concludes that they are as strongly held to the clay lattice as ice although the interlamellar water is layered rather than 3D-structured. This study led to the conclusion that the number of hydrates is related to the water content as shown in Table 2.3 for Na-montmorillonite with a dry density of 1200 kg/m³, corresponding to 1750 kg/m³ at complete water saturation.

2.2.5 *Sorptive properties of smectites*

Cations

The lattice charge deficit makes the smectites adsorb and exchange ions and inorganic and organic molecules. This is evident from the cation-exchange

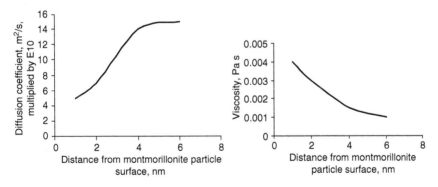

Figure 2.10 Diffusion coefficient and viscosity of porewater in montmorillonite calculated by use of molecular dynamics [10].

Table 2.3 Number of hydrate layers in compacted Na-montmorillonite with a dry density of 1200 kg/m³ as a function of the water content [5] (The clay was mixed with distilled water.)

Water content, % (by weight)	Number of hydrates
< 7	0
7–10	1
10–20	1–2
20–25	2
25–35	2–3
> 35	3

capacity (CEC) of smectites. Under a given set of conditions, different cations are not equally replaceable and do not have the same replacing power. The replacing power of some typical ions is shown as a lyotropic series as follows [11, 12]:

$$Na^+ < Li^+ < K^+ < Rb^+ < Cs^+ < Mg^{2+} Ca^{2+} < Ba^{2+} < Cu^{2+}$$
$$< Al^{3+} < Fe^{3+} < Th^{4+}$$

The replacement positions are to a very large extent dependent on the size of the hydrated cation. In heterovalent exchange, the selective preference for monovalent and divalent cations is dependent on the magnitude of the electric potential in the region where the greatest amount of cations is located. By and large, changes in the relative positions of the lyotropic series depend on the kind of clay and the ion being replaced. The number of exchangeable cations replaced depends on the concentration of ions in the replacing solution. The proportion of each exchangeable cation to the total CEC, as the outside concentration varies is determined by the exchange–equilibrium equations, the simplest useful one being the Gapon equation [12]:

$$\frac{M_e^{+m}}{N_e^{+n}} = K \frac{\left[M_o^{+m}\right]^{1/m}}{\left[N_o^{+n}\right]^{1/n}} \tag{2.2}$$

where (a) the superscripts m and n refer to the valence of the cations, (b) the subscripts e and o refer to the exchangeable and bulk solution ions and (c) the constant K is a function of specific cation adsorption and nature of the clay surface. K decreases in value as the surface density of charges increases. Na^+ versus Ca^{2+} represents a particularly important case of competition. Thus, it is a well-known fact that as the amount of exchangeable calcium on the clay mineral becomes less, it becomes more difficult to

Table 2.4 Typical CEC ranges for some important smectites [11]

Species	CEC, meq/100 g
Montmorillonite	80–150
Beidellite	80–135
Nontronite	60–120
Saponite	70–85

release. Sodium, on the other hand, tends to become easier to release as the degree of saturation with sodium ions becomes less.

Potassium is an exception. This is explained by the fact that its ionic diameter 2.66 Å is about the same as the diameter of the cavity in the oxygen layer. This allows the potassium ion to just fit into one of these cavities. As a consequence, the potassium ion is difficult to replace.

For other cations it is the size of the hydrated ion, rather than the size of the non-hydrated one that controls their replaceability. Thus, it appears that for ions of equal valence, those that are least hydrated have the greatest energy of replacement and are the most difficult to displace [11]. Li, although being a very small ion, is considered to be strongly hydrated and, therefore, to have a very large hydrated size. The low replacing power of Li^+ and its ready replaceability can be taken as a consequence of the large hydrated size, but there are in fact indications that Li^+ and Na^+ are only weakly hydrated in interlamellar positions.

Typical cation-exchange data of the most common smectite species are given in Table 2.4. For comparison, zeolites have a CEC in the range of 100–300 meq/100 g and are hence very effective cation catchers.

Anions

There are three types of anion exchange in smectites [11]:

1   Replacement of OH ions of clay–mineral surfaces. The extent of the exchange depends on the accessibility of the OH ions; those within the lattice are naturally not involved.
2   Anions that fit the geometry of the clay lattice, such as phosphate and arsenate, may be adsorbed by fitting onto the edges of the silica tetrahedral sheets and growing as extensions of these sheets [5, 6]. Other anions, such as sulphate and chloride, do not fit that of the silica tetrahedral sheets because of their geometry and do not become adsorbed.
3   Local charge deficiencies may form anion-exchange spots on basal plane surfaces.

The last mechanism is considered to contribute to the net anion-exchange capacity of smectites. The other two may be important in kaolinite but they

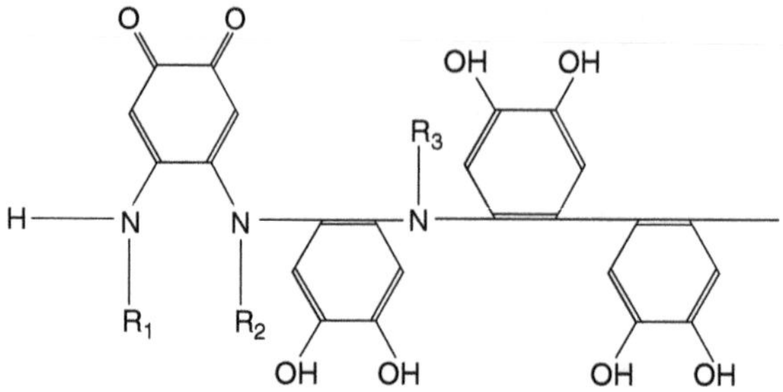

Figure 2.11 Soil-type humic acid forming polymers containing phenols and amino acids [13].

Table 2.5 CEC of montmorillonite determined with organic cations and barium, respectively [11]

Species	CEC, meq/100 g
Benzidine	91
p-Aminodimethylaniline	90
p-Phenylenediamine	86
α-Napthylamine	85
2.7-Diaminofluorence	95
Piperidine	90
Barium	90–94

are assumed to be relatively unimportant in smectite clays. The latter minerals commonly have an anion-exchange capacity of 5–10 meq/100 g but can be considerably higher for very fine-grained kaolinite and palygorskite.

Organic matter

Organic ions and molecules can enter the interlamellar space or be adsorbed on the edges and basal surfaces through hydrophilic groups or hydrogen bonds (Figure 2.11).

It appears from Table 2.5 that the CEC of the smectite species montmorillonite determined with certain organic cations is the same as when barium is adsorbed.

A recent research project has demonstrated the possibility for treatment of montmorillonite with organic substances to increase its anion-exchange capacity [7]. The study results showed that adding the chloride salt of the quaternary alkylammonium ion of hexadecylpyridinium ($HDPy^+$) can not only result in the complete exchange of the initially adsorbed cations, but also can be adsorbed in molecular form. We learn from this that iodide and

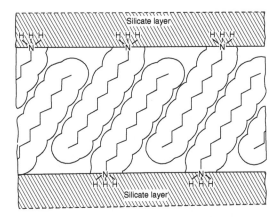

Figure 2.12 Generalized picture of positively charged organic ions or molecules in the interlamellar space of smectites [14].

pertechnetate sorption becomes very important while sorption of cesium and strontium decreases. For MX-80 bentonite, if such a treatment is applied, the sorption of iodide can be increased by several orders of magnitude. The organization of HDPy$^+$ in the interlamellar space of smectites can be of the type shown in Figure 2.12.

2.3   Non-expanding clay minerals in smectite clays

2.3.1   *Illite, chlorite and kaolinite particles*

No buffer material source contains solely smectite. Usually, both other clay minerals, rock forming minerals, amorphous constituents and organic matter are present in various amounts. The most common non-expanding clay minerals are illite (hydrous mica, hydro-mica), chlorite and kaolinite. Here, we will confine ourselves to a discussion of the hydration and sorption properties of these soils. They are commonly described with the following characteristic compositions [1, 14]:

- Illite: $M^I{}_ySi_{(8-y)}Al_yO_{20}(OH)_4(M^{III},M^{II})_4$
- Chlorite: $(Si,Al)_4O_{10}(Mg,Fe,Al)_6O_{10}(OH)_8$
- Kaolinite: $Si_2Al_2O_5(OH)_4$

where M^I is the monovalent cations (K$^+$ in illite), M^{II} is the Mg and Fe^{3+} and M^{III} is the Al and Fe^{2+}.

These compositions imply different intraparticle bond strengths and hence different particle dimensions. These impact on the specific surface area (SSA)

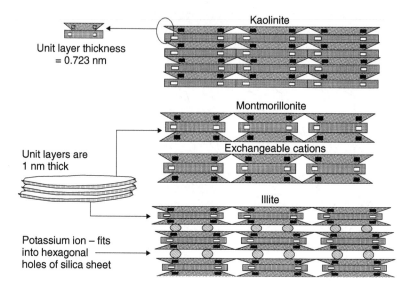

Figure 2.13 Kaolinite, montmorillonite and illite minerals, using the simplified crystal notation given previously. In illite and montmorillonite cation substitution of Si for Al is common in the SiO layer, and of Al for Fe and other cations in the Al–O–OH layer.

and the hydration potential. The crystal structures for the three clay mineral types are shown schematically in Figure 2.13.

2.3.2 Basic properties of illite, chlorite and kaolinite particles

Particles representing these clay minerals have hydrates on their basal planes to an extent that is determined by the crystal charge. The surface area of the particles is much smaller than that of smectite particles. This means that sorption is correspondingly smaller. The particles usually have a net negative charge at normal pH and a charge distribution that makes them form gels that are physically stable only at relatively high densities.

The non-expanding clay minerals commonly have a size smaller than 2 μm, which means that while the SSA is smaller than that of smectites, it is significant compared to non-clay minerals, and hence important to the water uptake and retention.

A typical size distribution histogram of illite particles in natural clays is shown in Figure 2.14. This histogram is also representative of chlorite particles. The corresponding SSA is in the range of 25–100 m²/g, or about 5–10% of that of the smectites and in the same order of magnitude as the external surface area of the latter clay type.

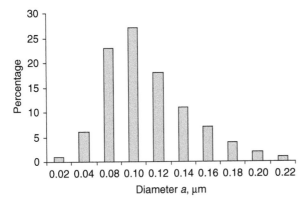

Figure 2.14 Typical size distribution of illite particles in the size range smaller than 0.22 μm. Natural sediment from 8 m depth.

The hydration potential is hence significantly lower than that of the smectites. The preceding notwithstanding, expressed in terms of the activity number a_c, which is a simple practical measure of the hydration potential, it is as high as 0.90 while Ca-montmorillonite has about $a_c = 1.5$ and Na-montmorillonite $a_c \sim 7$ [15].

The reason why the hydration of illite clay is not too different from that of Ca-montmorillonite is that the large size of the stacks of flakes in the latter clay is similar to the thickness of illite crystals. This results in a relatively small surface area. Interlamellar hydration is not extensive in any of these minerals. For comparison, the size distribution of particles smaller than about 0.3 μm of dispersed MX-80 clay is shown in Figure 2.15. The difference between the two clay types in the ultrafine size range is not very significant. However, the average length-to-thickness ratio of illite particles is commonly 5–25 and that of dispersed montmorillonite stacks of lamellae is at least 250. The reactive surface area of montmorillonite is therefore 10–50 times larger than that of illite particles. This explains the very significant difference in adsorption and retention of water of the two clay types.

The crystallites have a mineral structure similar to that of smectites. However, the interlamellar space is collapsed by the presence of potassium ions that hold the lamellae together. In addition to K^+ ions it is believed that H_3O^+ is present as well. These form hydrogen bonds with the neighbouring lamella. The charge distribution makes individual illite particles form stable gels but only at higher densities.

Illite

Figure 2.16 shows dispersed illite particles. Typically they have irregular edges and a stepwise reduced thickness close to them. Neoformed illite particles in weathering smectite may form thin regularly shaped laths.

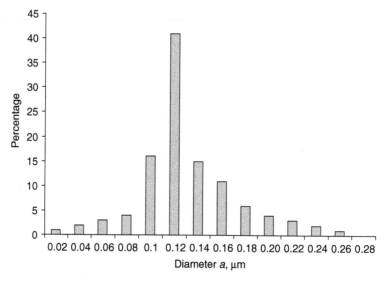

Figure 2.15 Size distribution (longest diameter) of MX-80 bentonite dispersed in distilled water.

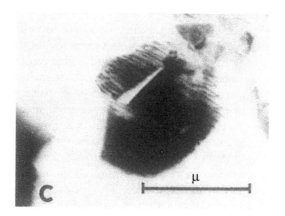

Figure 2.16 Dispersed illite particles photographed using a TEM. The edges are typically irregular and much thinner than the central parts of the particles. The saw-blade contour is caused by microtome cut of the lamellae.

Chlorite

Chlorite particles can have a particle size ranging from a few tenths of a micrometre to several millimetres depending on whether they emanate from weathered rock or result from neoformation in clay sediments. Chlorite never forms homogeneous zones of sedimented fine-grained material but is

Figure 2.17 Stacks of chlorite lamellae with lengths up to 20 μm.

a frequent component of glacial and postglacial clays in which they commonly make up 1–3% by weight.

The SSA of clay-sized chlorite particles is comparable to that of illite particles, and they are estimated to have similar hydration properties. Both are usually more or less weathered as revealed by XRD spectra, and they are occasionally somewhat expansive. Figure 2.17 shows hydrothermally formed chlorite particles in rock fractures. Such well-crystallized particles are rarely found in soils.

The relatively large SSA makes them form coherent clay gels in a manner similar to illites. They follow the same laws of colloid chemistry with respect to the impact of eleoctrolytes in the porewater as illites.

Kaolinite

The large diameter of the booklet-like kaolinite particles usually ranges from 0.3 to 4 μm while the thickness is 0.05–2 μm [15]. Numerous 7 Å thick lamellae form very thick, coherent stacks as in Figure 2.18. The particles of strongly cohering lamellae give kaolinite clays an SSA, that is, around 15 m²/g, and a low hydration capacity. In consequence, they therefore exhibit low plasticity and ductility with only a small difference between the Atterberg plastic and liquid limits. They behave in principle as very fine silt soil consisting of quartz grains.

Others

Other clay minerals that have some use in waste isolation are sepiolite and palygorskite (attapulgite). They are tube- or rod-shaped and usually

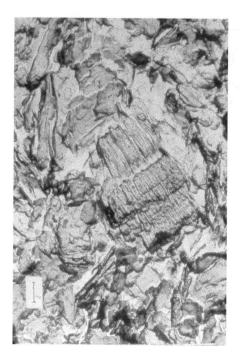

Figure 2.18 SEM picture of stacks of kaolinite particles.

possess a very high CEC. Natural deposits and material extracted for constructing waste isolating barriers have high hydraulic conductivity values that are too high to be of major interest, and are therefore not dealt with in this book.

2.3.3 Sorptive properties of non-expanding clay minerals

Cations

In *illite*, cation exchange takes place mainly on external surfaces, and because of the SSA and the surface charge characteristics, the CEC is relatively low. Since the grain size determines the SSA it is expected that very fine-grained illite will have significantly higher CEC than coarser illite material. This has been verified by various experiments [11], showing that the CEC is typically 15–20 meq/100 g for illite clay with a particle size of 0.1–1.0 μm. This increases to 30–40 meq/100 g when the particle size is smaller than 0.06 μm.

Chlorite exhibits properties that are similar to those of illite. Literature reports give values of CEC in the range of 10–40 meq/100 g [11]. *Kaolinite* is

a poor cation exchanger. Its CEC is generally in the range of 3–4 meq/100 g. This is attributed to its small SSA and to a low frequency of lattice substitutions. Grinding to a particle size of 0.1–0.05 μm increases its CEC to about 10 meq/100 g. The 'reactivity' and intercalation potential has recently been shown to increase with decreasing particle size [16].

Anions

Replacement of hydroxyls by anions is thought to be the main reason for anionic exchange. In montmorillonite, cation exchange is due mostly to lattice substitutions. The anion exchange capacity of montmorillonite depends on the frequency of exposed hydroxyls, which in turn is function of the ratio of edge to basal surfaces. It is hence a small fraction of the CEC of smectite, while, for kaolinite, with larger exposure of hydroxyls on the basal surfaces, the opposite conditions prevail [11]. Thus, the anion-exchange capacity can exceed 10 meq/100 g for very fine-grained powder. It is appreciably lower for illite and chlorite.

2.3.4 Mixed-layer minerals

Mixed-layer clays are obtained as an intermediate stage in the transformation of smectite to illite. This well-known process explains why sediments older than that of Ordovician age contain very little smectite. They are mentioned here since: (a) they are widely used for isolating waste in Germany (Friedland clay, FIM) and (b) they will be referred to in various discussions in subsequent chapters. In principle, they have a chlorite-type crystal structure and a swelling potential in the order of:

illite, chlorite $<$ mixed layer $\ll$ smectite

The swelling potential of these clays is higher than illite and chlorite, but markedly lower than smectites. Figure 2.19 shows the basic hydroxyl interlayer mineral where the octahedral interlayer joins the trioctahedral layers. This octahedral sheet, which forms the interlayer, has been called the interlayer hydroxide interlayer, a brucite sheet, a gibbsite sheet or an alumina sheet. When Al ions fill two thirds of the available positions in the interlayer hydroxide, the interlayer is known as the gibbsite layer. When Mg is in the octahedral sheet, the interlayer is known as the brucite sheet. The hydroxide interlayer differs from the regular octahedral sheet in that it does not have a plane of atoms which are shared with the adjacent tetrahedral sheet. Whilst Fe, Mn, Cr and Cu are sometimes found as part of the hydroxyl sheets, the more common hydroxyl sheets are gibbsite and brucite, that is, $Al(OH)_3$ and $Mg(OH)_2$.

A large number of mixed-layer minerals have been identified, especially smectite/illite types with their ratio indicating the degree of degradation of

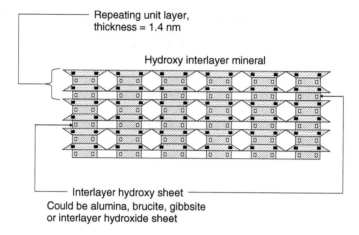

Figure 2.19 Basic hydroxy interlayer mineral with interlayer hydroxy sheets. Chlorite mineral is typical of this type of structure.

the smectite. Smectite/mica, smectite/kaolinite and smectite/vermiculite are also common in nature.

2.4 Rock-forming minerals in smectitic clays

2.4.1 *Typical constituents*

Rock-forming minerals are present in all sorts of natural smectitic clays. These are described in the general mineralogical literature – to which the interested reader is referred [14]. They may: (a) represent minerals formed in conjunction with the birth of smectite minerals in the evolution of bentonites for instance, or (b) be mixed in during transportation and sedimentation of smectite particles in allochthonous bentonites, or (c) be evolved from chemical degradation of ancient smectites. They may also represent unaltered grains in clay formed by hydrothermal processes in volcanic rock. As raw material with a higher smectite content than about 30%, smectite often constitutes a matrix in which rock-forming minerals and non-smectite clay minerals are distributed. They therefore do not have a very significant influence on the physical properties of the buffer. However, the chemical impact of certain minerals on canister corrosion, smectite conversion and cementation processes may be important, and the content of such minerals needs to be considered in the search of suitable canister-embedding clays. Such minerals are listed in Table 2.6.

It should be added that heavy minerals such as amphiboles and pyroxenes are apt to undergo degradation under hydrothermal conditions. These can be transformed to smectite – as documented by the numerous examples in

Table 2.6 Accessory minerals of major importance in buffers and
in the clay component of backfills (cf. Figure 1.2)

Mineral	Effect on buffer and canister
Feldspars	Source of Si and Al for formation of new minerals, colloids and cementing agents. K-feldspar provides K^+ for illitization
Sulphides	Source of SO_4^{-2} for attack on copper canister Source of SO_4^{-2} for formation of Ca and Na sulphates in the hottest part of the buffer
Carbonates	Source of Ca^{2+} for ion exchange of Na smectite to Ca form pH-enhancement

amphibolitic rock in Gothenburg, Sweden and many other areas. Naturally, this process improves the isolation potential of clays and backfills containing such minerals. Very often, mica grains and larger chlorite particles are more or less weathered as revealed by XRD spectra, and are occasionally somewhat expansive.

2.4.2 Hydration of rock-forming minerals

Unweathered rock-forming minerals have a low hydration potential and only one hydrate layer is formed on their surfaces. However, although making up only a small fraction of the total mass, the monolayer on quartz and feldspar particles is of importance in preparing mixtures of clay and ballast (aggregates) for backfilling of tunnels and shafts in radioactive waste repositories. This is because the usually low 'optimum' water content of such mixture depends very much on the moisture provided by the rock-forming minerals. The organization of water molecules at the mineral surfaces is believed to be of low order and the water film does not contribute to the frictional resistance to shearing.

2.4.3 Ion exchange and sorption of rock-forming minerals

The cation and anion-exchange capacities of rock-forming minerals is negligible. However, as constituents in ancient sediments they are often partly weathered and therefore have some ability to sorb cations.

2.5 Inorganic amorphous matter in smectitic clays

Dissolved mineral particles will release free silica, aluminium and magnesium in the porewater. Some of these elements spontaneously rearrange to

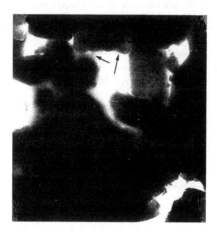

Figure 2.20 TEM micrograph of amorphous substance precipitated on carbonate particles cementing them together. Edge length of micrograph is 10 μm.

form inorganic colloids such as Fe and Al compounds (sesquioxides), which are known to be mobile in shallow natural soils. It is doubtful that they can move in the very narrow and tortuous pathways of highly compacted bentonite. Iron and sulfur combine to form colloids with similar mobility [14]. The latter reactions are very redox-sensitive, however, and polysulphide complexes may form instead. Formation of inorganic amorphous matter in smectite buffer clay under hydrothermal conditions may be of great practical importance since it may serve as cementing agent. This will in effect reduce the expandability of the clay. It may also have a positive effect by clogging voids and fractures in the rock surrounding the buffer – assuming that it can migrate and enter such openings. Figure 2.20 illustrates how amorphous coatings of carbonate particles can appear. The lack of crystallinity is easily identified using electron diffraction techniques.

2.6   Organic material in smectitic clays

The various types of organic species in smectitic clay range from millimetre-sized tissues emanating from plants, fungal and animal activities, to organic molecules representing ultimate degradation products of organic life. Positively charged organic molecules can enter the interlamellar space in smectites and organic compounds can link particles together both by physical action and colloidal forces (Figure 2.21). A problem that can arise in the practical use of smectite clays is that mobile organic colloids can be formed and adsorb radionuclides. These in turn may be transported to the biosphere [1, 16].

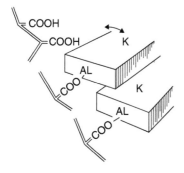

Figure 2.21 Bonding mechanism of organic molecules and clay minerals.

Bacteria contained in bentonite raw material that is used for preparation of buffers and backfills can act as mobile colloidal particles. They may multiply and become numerous in soft smectite gels corresponding to the clay component in poorly compacted backfills in repositories. However, recent theoretical considerations and laboratory investigations have shown that bacteria of the kind that represent a potential risk of transporting radionuclides do not survive in highly compacted bentonite [17]. This is because both access to free water is very limited and because the interconnection and sizes of the voids are limited – with many voids smaller than the bacteria.

The organic substance in bentonites can be effectively removed by heating the air-dry powder at about 400°C for 1 hour without significant alteration of the smectite minerals [1]. The matter of acceptable content of organics is therefore not very critical although it has an economic impact. For the Swedish organization responsible for radioactive waste handling and disposal (SKB), an organic content of 2000 ppm is considered to be acceptable.

A most important microstructural feature is the presence of gas emanating from organic constituents. Figure 2.22 shows a rounded void filled with gas emanating from organic matter in the clay.

2.7   Origin and occurrence of commercially exploited smectitic clays

2.7.1   *General*

Smectites stem from spontaneous nucleation and growth of crystals in saline water with the glassy components of volcanic ash as source material. They can also be derived from weathering of feldspars and heavy minerals in hydrothermally affected rock. The first-mentioned material is true bentonite, which forms beds of different thickness in many parts of the

Figure 2.22 TEM micrograph gas-filled void in clay matrix. The gas emanated
         from the nodular assembly at the upper left of the void and the
         broom-like aggregate below the lower end of the void.

world. Smectites derived as weathering products are confined to rock
masses with a structure that has been sufficiently pervious to allow solutions through – such as fracture-rich zones in crystalline rock and porous
rock of sedimentary or volcanic origin.

The following summary is a very short overview to highlight the wide-
spread occurrence of smectitic material. We will pay more attention to some
North American, Japanese and European bentonites in our discussions
since they will be referred to in later discussions in the book.

2.7.2   Bentonites

Many of the major smectite sources, particularly bentonites, are exploited
for industrial use and are candidates for manufacturing clay seals in repositories. The well-known examples are the bentonites in Wyoming, and southern Europe. Many of these were formed during a marine episode in
Cretaceous time where extensive volcanic activity with large quantities of
ash were formed and blown by the prevailing winds. The result was a series
of ash deposits in seawater and brackish estuaries. The deposited ashes were
subsequently buried by marine or land-derived materials. With time and further burial by other overburden material, consolidation of the deposited
ashes resulted. Many of these ancient ash layers later altered to bentonite.

In the Black Hills area in Wyoming the bentonite beds have a thickness of a few centimetres to a few metres and appear in a series of marine shales, marls and sandstones. The Wyoming bentonites have been exploited by the American Colloid Co. for decades for production of various commercial clay materials such as the MX-80 clay. This is a smectite-rich clay with a special grading. It has a particularly good reputation amongst the large number of commercial smectitic clays because it has Na as the major adsorbed cation. It exhibits good colloidal, plastic and bonding properties.

European bentonites of the same age almost always have Ca as the major cation and are converted to Na form by soda treatment. This results in a slightly higher calcite content than the natural Na-dominated bentonites. Commercial data on clay manufacturers and export and import of bentonite materials are available from various sources, and Table 2.7 is a compilation thereof.

North America

The most widely used and mined North American deposits of sodium bentonite were formed during the Cretaceous age through a combination of geological activity and environmental conditions that allowed initiation of bentonite formation. Central North America had been inundated by a

Table 2.7 World production (P) and exports (E) of bentonite per 1997

Country	Production, kilotons	Export, kilotons
USA	6390	1007
Germany	1111	73
Greece	950	12
Spain[a]	846	603
Italy	543	182
Turkey	531	62
Japan	496	2
India	367	174
Korea	167	6
Mexico	163	13
UK	135	158
Argentina	132	35
Brazil	110	—
Czech Republic	110	21
Cyprus	100	52
Senegal	100	—
Others	691	404
Total	12 942	2804

Note
a incl. sepiolite.

shallow brackish sea that varied with time, especially in the northwestern areas (now Alberta and Saskatchewan). Contemporaneous with this marine episode was extensive volcanic activity in the Western Cordillera. Vast quantities of ash were formed and blown east by the prevailing winds. The result was a series of ash deposits on the inland sea. They gradually settled and became buried by marine or land-derived materials. With time, these deposits were deeply buried and consolidated by the more recent sediments and many of these ancient ash layers later altered to bentonite.

The bentonite beds have a character that depends on the initial ash composition and granularity, as well as on the pore-fluid chemistry of the deposits. Those in the North and West tend to be more 'ashy' with a lower smectite content due to their proximity to the volcanic sources of the ash, or to the inclusion of terrestrial material carried into the deposits during sea level fluctuations. The deposits in the central basin tend to be more homogeneous, perhaps as the result of calmer central waters and finer textured material deposited farther from the volcanic ash sources.

The United States has large bentonite resources of the Upper Cretaceous and Tertiary ages. The principal producing areas are the Black Hills region of South Dakota, Wyoming and Montana, and the Gulf region, especially in Texas and Mississippi. In the Black Hills area the bentonite beds have a thickness of a few centimetres to a few metres and appear in a series of marine shales, marls and sandstones. It is generally assumed that the bentonites in this region were formed by *in-situ* alteration of volcanic ash, believed to have been rhyolitic and stemming from a western source. The best producing horizon has been the 'Clay Spur' bed forming the top of the Mowry shale. The Wyoming bentonite has a particularly good reputation amongst the large number of various smectitic clays because it has good colloidal, plastic and bonding properties.

Very often the beds cover more or less silicified (silicitated) strata, indicating release of silica from the smectite in conjunction with its formation. At the outcrop, bentonite is often light yellow or green whilst it is bluish deeper down. The colour shift is the result of oxidation of iron or leaching replacement of exchangeable sodium by calcium. However, sodium is commonly the dominant adsorbed cation in the bentonite from Wyoming and South Dakota. In the Gulf area, most bentonites have calcium as the major adsorbed cation.

The important question as to what the controlling factors really are for the type of exchangeable cation in natural bentonites has no definite answer. Marine and lacustrine formations may have either sodium or calcium as dominating cation. This would suggest that the composition of the ash is a determinant.

Canada is rich in bentonite, particularly in the Prairie Provinces (Manitoba, Saskatchewan and Alberta). The bentonites, which are similar to those in Wyoming and South Dakota, are of Cretaceous and lower Tertiary ages. They occur in marine shales and limestones and their thickness ranges between a centimetre and a few metres. Alteration of volcanic ash is the generally accepted mode of formation. Most of the Prairie bentonites have calcium as major exchangeable cation but sodium-dominant clays occur as well. Cristobalite is a common accessory mineral. This may explain the slight cementation that is characteristic of many of the Prairie bentonites. In fact, they are often shaly and somewhat brittle and fractured.

South America

South America has bentonites in Argentina, Peru, Uruguay and Brazil. Clay originating from altered tuff is found in older, that is, Triassic, formations (Mendoza) in Argentina. Cretaceous smectite-rich bentonites formed by ash fall in lacustrine environment occur in Brazil in the Ponta Alfa area of Minas Gerais, and Permian bentonite beds, up to 3.3 m thick, have been found in the Estrada Nova formation at Acegua in Rio Grande do Sul. Both types are rich in montmorillonite, while in other Brazilian bentonites, the smectite is diluted with substantial amounts of other minerals.

Central America

Mexico is rich in bentonites in the volcanic region. They stem from volcanic ash and have calcium as the major adsorbed cation, although natural Na-smectites of high quality have been reported as well, unfortunately not in written form (American Colloid Co.). Chalcedony overlying and underlying certain beds may be related to heating, in conjunction with hydrothermal alteration of the ash. The extension and quality of the bentonite sources are poorly known.

Africa

Bentonites exist in many areas in Algeria and Morocco. Some of them originate from volcanic ash, such as those in the Chelif River area in Algeria whilst others have been formed by hydrothermal alteration of rhyolite rock. The bentonites formed by ash fall vary in thickness from about 10 cm to around 5 m. They represent Cretaceous and Tertiary sediments with interbedded pyroclastics and lava flow. The complex stratification indicates variation in homogeneity and smectite content. Some of them have a high smectite content with sodium as the major adsorbed cation and with no enrichment of silica. A number of these bentonites have been extensively utilized commercially.

In Morocco, bentonites of Miocene age are common, often in the form of layers with a thickness of a few centimetres. They stem from ash fall. An area known to have smectite-rich bentonites of high quality is Camp-Berteaux in the Taourint region. The smectites from this area serve as one of the French reference montmorillonite minerals.

In South Africa, bentonite beds of Triassic age occur with a thickness of more than 2 m in the Karoo system. These beds, which have calcium as the major exchangeable cation and do not originate from ash fall, are composed of very pure montmorillonite. In Mozambique, thick bentonite masses, formed by hydrothermal alteration of perlite occur, but they contain a significant proportion of unaltered rock fragments and cristobalite.

Asia

India has relatively pure smectite clays of the lower Tertiary age. They form large deposits with more than 3 m thickness and are thought to stem from *in-situ* weathering of igneous and metamorphic rocks followed by erosion, transportation and deposition in the Barmer Embayment. One metre thick bentonite beds have been found in the Jamash and Kashmir areas. They seem to have been formed by alteration of volcanic glass in ash.

Japan

Clay formation has been extensively studied in Japan. These studies have led to a considerable deepening of the understanding of the mechanisms involved in bentonite formations. An important conclusion from this is that it is unlikely that weathering processes were important in the formation of such bentonites in Japan. Studies show that weathering tends to produce a suite of clay minerals rather than a single one, and that other clay minerals rather than smectite are likely to be abundant. In most of the bentonites originating from hydrothermal influence on rhyolitic rock, there are frequent quartz phenocrysts and a considerable quantity of plagioclase and biotite. The smectite content is hence not very high.

Bentonites formed from volcanic ash and by *in-situ* alteration through hydrothermal processes in Tertiary liparites and liparitic tuffs are present in several sites in Honshu and Hokkaido. At Yamagata and in the Tsukinuno area on the main island, the deposits were formed in Early Miocene mostly by hydrothermal processes that produced large clay bodies which are irregular in shape. Depending on the composition of the parent rock, zeolites have also been formed in large amounts. In the Kwanto district of the Gumwa Prefecture there are several beds with a total thickness of up to 5 m, interbedded with mudstones and hence of heterogeneous and relatively poor quality.

A general conclusion is that most of the clays originated from volcanic ash irrespective of the formation process. Since the volcanic activity of Japan has been very significant from the Neogene epoch, the amounts of clay are very large. In particular, the submarine sedimentation of the Miocene dacite pumice was very rich and most of the deposits show huge masses of montmorillonite, zeolite and opal. Most of the bentonites that are being exploited commercially are altered tuffs. At Tsukinuno, the geological profile is as shown in Table 2.8.

Table 2.8 Geological profile in the Tsukinuno Area (Tertiary age)

Formation and age	Unit	Thickness (m)	Richness
Nukumi F	Gray tuffaceous sandstone	80	None
	Brown pumiceous tuff	30	+
	Dark brown hard mudstone	20	
	Gray pumiceous tuff	20	+
Tsukinono F	Dark brown hard mudstone	290	+
	Brown hard shale	100	+ + +
	Pale green sandy tuff	100	+

Note
+ means fairly bentonite-rich, + + + very rich.

Europe

European bentonites are well known and easily available in Spain, Germany, Denmark, Hungary, Czech Republic, Italy, Greece and Bulgaria. We will confine ourselves here to mention those in Germany, Italy, Spain, Czech Republic, Greece and Sweden. Those from Germany, Spain and the Czech Republic have all been examined in international research work on buffers and backfills as will be discussed in later chapters.

Germany

There are important sources in Germany of commercially exploitable bentonites, generally emanating from *in-situ* alteration of acid vitreous tuff. At Moosburg in Bavaria, bentonite with calcium as the major exchangeable cation forms up to 3 m thick beds in a generally marine section of marls and tuffaceous sands. The purity varies from relatively non-contaminated montmorillonite to material with a large proportion of kaolinite and illite. Typically, there is a distinct boundary between an upper oxidized (yellow) and a lower blue-coloured part in one of the pit areas.

In the Neubrandenburg region of northern Germany, there are huge quantities of smectitic clay that are exploited for isolation of shallow waste piles. This material is presently being considered as a possible material for backfilling of drifts and shafts in repositories for radioactive and hazardous chemical waste. The clay, which is of Tertiary age, has an average content of expandable minerals of 50–60%, half of which is montmorillonite and the rest being mixed-layer minerals. The clay is called *Friedland Ton*. This clay will be used as an example (for discussion in subsequent chapters) of

Figure 2.23 View of open pit mining of Friedland Ton (FIM GmbH). The pit is
about 100 m deep (see Colour Plate VI).

commercially available relatively smectite-poor European clays that are
potentially useful for repository sealing (Figure 2.23).

Italy

Bentonites are abundant, particularly in the islands of Ponza, Sardinia and
Sicily. Most of them originate from alteration of basic rocks, some due to
the effect of gases left in intrusions of magma whilst others appear to be of
sedimentary origin. One case of the latter type is concluded to be the
presently mined bentonite at Busachi in central Sardinia. Here, molten rhyolite is assumed to have moved in, over a more than 7 m thick reddish,
rather homogeneous bentonite mass. This has produced sintering at the
contact, but does not appear to have caused much changes at distances of
more than 2 m from the hot contact. This and similar sites, including the
Uri, Oristano and Cagliari regions, represent rich potential bentonite
sources.

At Cagliari, the bentonite reserves represent tens of millions of cubic
metres with up to 10 m thick bentonite beds. At the Pedra de Fogu area
the bentonite beds are up to 25 m thick. At Cagliari the bentonite is a
Miocenic alteration product of a more than 13 m thick fine pyroclastic
stratum, whilst at Pedra de Fogu it stems from *in situ* alteration of
rhyolitic rock.

Spain

In recent years, systematic prospecting has given a good overview of the Spanish bentonite resources. Three areas have turned out to be of particular practical importance. They have all been formed by alteration of Tertiary rhyolite-dacites through hydrothermal processes or percolation of meteoric water involving removal of some silica and iron, and the incorporation of aluminium and magnesium. Concentrations of chromium, nickel and cobalt have caused colouring of some of the bentonites.

The Madrid Basin, part of the Tajo Basin, lies in the southern edge of the Sistema Central. The Tajo Basin is a Tertiary depression, located in the middle of the Iberian Peninsula, which became filled with continental sediments, ranging from fluviatile to lacustrine-palustrine environment-type deposits. A major characteristic of these deposits is a lateral change of facies from detrital to chemical. Clay materials of saponite type are commonly found in the transition zones and they represent about $500\,000$ m^3. Three different units have been established: (a) a lower unit composed mainly of evaporites, (b) an intermediate unit made up of clays, sandstones, limestones and chert, and (c) an upper unit of essentially detrital materials, associated with tectonic phases that occurred during the Miocene. The intermediate unit is of particular interest and is made up of materials from lacustrine and fluviatile environments with some transition zones. Predominating in the transition zone are clay materials associated with silty sands. Biotitic sepiolite deposits, exploited in the Madrid Basin, are found within these clay formations.

Tests performed in the programme of the Spanish organization ENRESA responsible for disposal of Spanish HLW have shown that the Cerro del Aguila bentonites with a smectite content of more than 85% have the best characteristics of three deposits identified in the Madrid Basin. Available field data from borings indicate that the potential reserves and an 'easy to mine' configuration of the deposits meet the set requirements.

The main Spanish bentonite deposits of hydrothermal origin are located in Cabo de Gata (Almeria, southeastern Spain). The volcanic activity in this area gave rise to calc-alkaline rocks, from basaltic andensites to rhyolites in composition. The volcanic emissions varied from massive lava flows to different types of pyroclastic products. The rocks were heavily fractured and have undergone comprehensive physical and chemical alteration. Most of the pyroclastic materials have experienced hydrothermal alteration. Acid solutions gave rise to the formation of alunite, jarosite and kaolinite. However, the most extended alteration resulted in the formation of the thick bentonite deposits spread all over the Cabo de Gata region. The bentonites in Serrata de Nijar cover an area of about 11×1.5 km^2 and extend to at least 20 m depth. The smectite content is in the interval 85–95%, the major species being montmorillonite. Other sites are the montmorillonite-rich zones of Los Trancos and Escullos.

The process of alteration of the volcanic materials to bentonite took place with a large loss of matter, in particular silica and alkaline elements, accounting for over 50% of the fresh rock mass. The accompanying decrease in volume gave rise to some morphologic readjustments in the altered pyroclastic bed cover. As a rule the hydrothermal solution supplied Mg to the initial fresh rock.

The meteoric waters, heated during their underground southward movement from the metamorphic northern ranges altered the pyroclastic beds into bentonite. Since the hydrolysis reaction of these materials is exothermic, the solutions did not need to reach great depths or high temperature for the bodies to be heated up. These smectites belong to the montmorillonite–beidellite series and their compositions vary gradually with the degree of alteration and the chemical composition of the parent materials. Thus, beidellites correspond to very early alteration stages, especially in acid rocks. The colour of the bentonite depends on the content in trace elements of the first transition series; all shades of green, blue, red and white are present. The bentonite deposits are numerous and important all over the region and reserves are estimated to be tens of millions of cubic metres.

Czech Republic

The Bohemian Massif in the Czech Republic is estimated to contain at least 280 million (metric) tons of high-grade bentonites representing 16 sites – of which 3 are being mined. One of them is the Zelena clay quarry in the upper Miocene Cheb basin, which holds up to 20 m thick smectitic layers. Both Ca-dominated clay requiring soda activation and Na-dominated clays have been identified and exploited.

Greece

The volcanic islands of Greece were formed by transgressive sedimentation on a crystalline base, a dominant part of the island being covered with sediments of volcanic origin deposited in late-Tertiary and early-Quaternary times. Alluvial sediments are found in tectonically generated depressions. Volcanic activity was initiated in the Pliocene age forming tuffs and tuffites and proceeded in five phases. This has resulted in a considerably complex formation. The geothermal conditions are still very obvious as demonstrated by the fact that water vapour with 300°C temperature is contained in pores and fractures at less than 1 km depth. The island of Milos represents a major bentonite resource. The geology of Milos has been described in detail with respect to the presence and extension of smectitic clays.

There are several mineralogically and chemically different clay beds, the ones of major practical interest being formed by *in-situ* alteration of andesitic pyroclastic rock characterized by high contents of volcanic glass, and plagioclase that were transformed to montmorillonite in many places.

The main clay beds are in one area, Agrilies. These consist of iron-poor smectites with high quartz and mica contents. Xenolites in the original rock are largely preserved here. In the Ano Komia area, bentonites with little smectite and significant amounts of cristobalite and tridymite were formed from pyroclastic rock. In certain areas, particularly in shallow positions and in shear zones, hydrothermal solutions have percolated and degraded the smectites. In the area Aspro Chorio, where smectite-rich bentonites are available in very large quantities, the processes involved in the transformation of pyroclastic rock and the evolution of smectites have been studied in detail. It is claimed that percolation of meteoric water of relatively porous rock was a major factor in the smectite formation in this and other areas.

Eastern Europe

Large resources of smectite clay have been identified in Ukraine and other areas in the former Soviet Union. Montmorillonite and saponite are major smectites, often appearing together with palygorskite (attapulgite), which has a very high CEC but also a high hydraulic conductivity. This makes the material useless as a buffer or barrier material for waste isolation. The very comprehensive Russian research on clay behaviour, particularly the fluidity and hardening of thixotropic drilling muds, has been based on Ukrainian smectite and palygorskite [18], and the contribution to the present understanding of the gas tightness of ancient clay sediments is substantial [19]. This will be discussed further in this book.

2.7.3 *Potential smectite clay resources*

The formations described in this chapter represent very large amounts of clay that can be used as raw material for enrichment of smectite minerals or for use as it is. Complete data of the quantity of smectitic clay is usually not available but approximate figures can be obtained from certain companies or derived from stratigraphic data. The total known resources on land of bentonitic materials in the countries referred to previously is estimated to be about 10^{11}–10^{12} m^3, but the actual figure may be much higher [14].

The outcropping Black Hill bentonites cover an area of at least 500 km^2 corresponding to at least E10 m^3 of bentonitic material. The bentonite deposits in the Gulf States probably represent about the same quantity. Japanese bentonites also represent very large volumes. Thus, in the Prefecture of Gumma the bentonite zone is about 25 km long and 2 km wide, corresponding to at least 10^8 m^3 bentonite-holding material. The total amount of exploitable bentonitic material in Japan can be estimated at 10^{10} m^3.

The total European resources, excepting Russia, is estimated at more than 10^{10} m^3. In Italy the Sardinian resources of bentonitic clay represent no less than 10^9 m^3 and the entire amount of such clays in the whole of Italy probably exceeds 10^9 m^3. The Greek and Spanish resources are estimated to be on the same order of magnitude.

2.8   References

1. Pusch, R., 1994. *Waste Disposal in Rock. Developments in Geotechnical Engineering, 76.* Elsevier Scientific Publishing Company, Amsterdam, Oxford and New York.
2. Kehres, A., 1983. Isothermes de deshydration des argilles, energies d'hydration–Diagrammes de pores surface internes et externes. Doctoral Thesis Universite Paul Sabatier de Toulouse.
3. Forslind, E. and Jacobsson, A., 1972. 'Clay-water systems', in *Water, a Comprehensive Treatise*, F. Franks (ed.), Plenum Press, New York.
4. Sposito, G., 1984. *The Surface Chemistry of Soils.* Oxford University Press, New York.
5. Pusch, R. and Karnland, O., 1988. Hydrothermal effects on montmorillonite. A preliminary study. SKB Technical Report TR 88–15, SKB, Stockholm.
6. Pusch, R., 1993. Evolution of models for conversion of smectite to non-expandable minerals. SKB Technical Report TR 93-33, SKB, Stockholm.
7. Pusch, R., Muurinen, A., Lehikoinen, J., Bors, J. and Eriksen, T., 1999. 'Microstructural and chemical parametres of bentonite as determinants of waste isolation efficiency'. Final Report EC project Contr. F14W-CT95-0012. European Commission, Brussels.
8. Colten, V. A., 1986. 'Hydration states of smectite in NaCl brines at elevated pressures and temperatures'. *Clays and Clay Minerals*, Vol. 34: 385–389.
9. Nakano, M., Fuji, K., Hara, K., Hasegawa, H. and Fujita, T., 1998. 'Changes of the stacked structure of montmorillonite in unsaturated bentonite with compression: Estimation based on calculated chemical potential of water'. In *Proceedings of the Workshop on Microstructural Modelling of Natural and Artificially prepared Clay Soils with Special Emphasis on the Use of Clays for Waste Isolation*, Roland Pusch (ed.), Clay Technology AB, Geodevelopment AB, SKB, POSIVA, EC.
10. Ichikawa, Y., Kawamura, K., Nakano, M., Kitayama, K. and Kawamura, H., 1998. Unified molecular dynamics/homogenization analysis. In the *Proceedings on International High-level Radioactive Waste Management Conference*, Las Vegas, NV.
11. Grim, R. E. and Gueven, N., 1978. *Bentonites. Developments in Sedimentology 24.* Elsevier Scientific Publishing Company, Amsterdam, Oxford and New York.
12. Yong, R.N., 2001. *Geoenvironmental Engineering: Contaminated Soils, Pollutant Fate, and Mitigation.* CRC Press, Boca Raton, FL, 307pp.
13. Pusch, R., 1973. Influence of organic matter on the geotechnical properties of clays. Document D11:1973, Swedish National Council for Building Research, Stockholm.
14. Gast, R. G., 1977. 'Surface and colloid chemistry', in *Minerals in Soil Environment*, R.C. Dinauer (ed.), Soil Science Society of America, Madison, WI.
15. Pusch, R., 2002. The Buffer and Backfill Handbook, Part 2. SKB Technical Report TR-02-12, SKB, Stockholm.

16. Franco, F. and Ruiz Cruz, M. D., 2004. 'Factors influencing the intercalation degree ("reactivity") of kaolin minerals with potassium acetate, formamide, dimethylsulphoxide and hydrazine'. *Clay Minerals*, Vol. 39, No. 2: 193–205.
17. Borchardt, R., 1977. 'Montmorillonite and other smectite minerals', Ch. 9 in *Minerals in Soil Environments*, Soil Science Society of America, Madison, WI.
18. Pusch, R., 1999. Mobility and survival of sulphate-reducing bacteria in compacted and fully water saturated bentonite – microstructural aspects. Technical Report TR-99-30, SKB, Stockholm.
19. Ovcharenko, F. D., Nichiporenko, S. P., Kruglitskii, N. N. and Tretinnik, V. Yu, 1967. 'Investigation of the physico-chemical mechanics of clay-mineral dispersions', P.A. Rehbinder (ed.), *Academy of Sciences*, Ukrainian SSR. Transl. by Israel programme for Scientific Translations, Jerusalem, Israel.

3 Microstructure of natural smectite clay

3.1 Objective

Why are natural smectite clays of interest with respect to its role and utility as a significant component in engineered barrier systems (EBS) – in the field of geoenvironmental engineering, and particularly in environmental geotechnics? There are at least three reasons:

1. They can be used directly as buffer and liner systems, or in their natural form to effectively seal off repositories for hazardous wastes located in or below such clay layers.
2. The properties and characteristics of the microstructural units (MUs) of natural smectite deposits used to manufacture and construct artificial barriers significantly affect their capability to isolate and seal the repositories.
3. The knowledge obtained in respect to the degradation that ancient smectitic clay strata have undergone provides us with an indication of how smectite clay used in EBS will change or transform in the longer term – in the order of tens to hundreds of thousands of years.

The first issue is well illustrated by some repository concepts that have been proposed for deep disposal of highly radioactive waste. It has been suggested that locating a repository in mechanically stable crystalline rock at 600 m depth, below a series of extremely tight bentonite layers at the southern end of the Swedish island Gotland would offer a perfectly tight seal against contamination of the biosphere by radionuclides [1]. The same first issue is also important in respect to the understanding of gas tightness and how this can be moderated for landfill top cover [2]. The second and third issues provide one with a basis for predicting the long-term performance of the clay and its effectiveness in the isolation of hazardous waste – as, for example in the repository scenario of 'buffer embedding canisters with highly radioactive waste [3]'.

3.2 Definitions

There is no universally accepted agreement on the concept and definition of the microstructure of soils. That being said, there exists a common interpretation which includes the geometry and properties of the system of particles and voids in the description or characterization of the microstructure of a soil. Some investigators also include the interparticle force fields in this concept, implying that they are determinants of the particle arrangement. To distinguish between the geometrical interpretation of soil structure and the one which includes the interparticle forces, the terms *soil fabric* and *soil structure* have been used – with the former referring to the geometrical concept and the latter including the interactions between particles.

The difference between *soil fabric* and *soil structure* is more than a terminology difference. Although the description of the arrangement of particles and aggregation of particles (fabric) is important, it is the interactions between these particles and their aggregate groups (i.e. microstructural units, MUs) that provide the soil with its characteristics and properties. In the hierarchy of structural components, one begins at the bottom with single discrete particles. In the case of phyllosilicates, that is, layer lattice minerals, these can range from the kaolinites to the stacked lamellae minerals typified by the smectites. As noted in Chapter 2, the smectites are known to be stacked lamellar (also called stacked layer) sheets with interactive properties between the lamellae, whereas the kaolinites, do not (see Figure 2.12).

The level above the discrete single particles is the MU. This is formed naturally because the bonding forces and energies of attraction between particles are considerably larger than the energies of repulsion. These will be described in a later section. The term microstructural unit (MU) is used as a general term to include such designations as peds, flocs, clusters, domains, aggregate groups, crumbs, etc. All of these terms mean the agglomeration or aggregation of particles resulting in a grouping of particles that act mainly as a single unit. The microstructure of a soil is seen to be defined by the properties of individual MUs and the manner in which they are bonded and interact. This matter will be discussed first in this chapter. Proceeding further to the description of the arrangement of particles, there are various parameters for characterizing the microstructure. This is particularly true with regard to the spectrum of voids that will be discussed in the later chapters. A necessary prerequisite for quantification of microstructural features is to prepare images of them. Techniques for making it possible to produce digital microstructural pictures that can be analysed by modern techniques will be described here as well.

3.3   Microstructural evolution of sediments

3.3.1   *General*

The role and influence of soil structure on the properties and performance of clays can be studied from many different perspectives, depending upon end-purpose use or goals. The importance of the microstructural components and their contribution to the overall structure of a soil has long been recognized in the field of Soil Science because of their need to deal with such problems as soil moisture movement, soil tillage, water-holding capacity of soil, soil–water potentials and water uptake of plants. In the field of Soil Engineering, attention has historically been directed more towards the physical and mechanical performance of soils. Because of the apparent applicability of continuum mechanics as a tool for analysis of soil performance, soils were initially treated as 'uniform and homogeneous' materials, that is, a continuum. As time progressed, it became clear that certain properties could not sustain analysis by treating soils as a continuum. Prominent amongst these are such 'irregular' properties and performance as anisotropy, rheotropic performance, preconsolidated and overconsolidated clays.

The study of Soil Behaviour, which came into prominence in the 1950s and beyond, provided the impetus to investigate and determine the role of soil microstructure in the control of soil properties, behaviour and performance. For many soil engineering applications, consideration of the soil as a continuum is appropriate. This is particularly true when: (a) the testing procedures used are designed to measure and/or determine the macroscopic performance of the soil, (b) the soil property testing-analytical tool used in laboratory or field data reduction relies on data-reduction models developed for uniform and homogeneous media and (c) the analytical tools used for assessment of soil performance are based on continuum mechanics principles. Problems arise in prediction of soil behaviour when the soils under consideration do not obey performance characteristics consistent with the principles of the mechanics of continua. Two particular reasons exist: (1) The soils are heterogeneous and structured and therefore require consideration of the structure of the medium. The case of sensitive marine clays is a good example; (2) The behaviour of the soil responds to both gravitational and molecular forces, and hence requires analyses that incorporate intermolecular forces in the overall analytical package. This is particularly true for clay soils with active clay minerals, that is, clay minerals with active and reactive particle surfaces. Smectite is a good example of this type of clays.

In general, there exist several types of soil-particle structural units in a clay soil. These range from individual mineral particles to flocs, clusters and peds, known as MUs. These MUs (flocs, clusters and peds) are aggregations of particles and are the backbone of the macrostructure of clay soils – as shown schematically in Figure 3.1. Since particle sizes are in the micron range, and since the specific surface areas (SSAs) of these particles and structural units are in the order of ten to hundreds of square metres per gram of soil, it is evident that intermolecular forces will dominate the interactions between these particles and MUs. In the case of smectites, we have seen in Section 2.2.1 that these consist of interacting layer-lattice structural sheets identified as *stacked lamellae* or *stacked layers*.

The intralamellar interactions (within the stacked lamellae) add to the complex interactions. The combined interlamellar and intralamellar

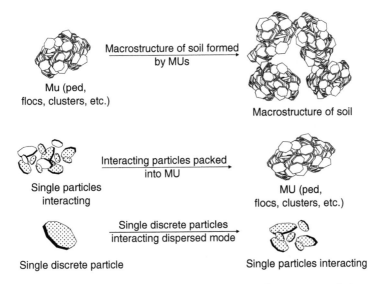

Figure 3.1 Hierarchic representation of soil structure beginning with interacting discrete particles and progressing onward to microstructural units and macrostructure of soil.

interactions in smectites in the MUs, together with interactions between such microstructures are generally identified as physico-chemical interactions. These physico-chemical interactions produce the observed characteristics, properties and response performance of such soils.

3.3.2 Structure and particle interaction

In a natural setting, the development of the structure of soil (macro and micro) is a function of the physico-chemical and biogeochemical interactions in the soil-particle-water system. This involves the surfaces of the interacting soil solids and porewater. Interactions between particles, that is, particle-to-particle interactions, are common phenomena in soil suspensions.

Discharge waste tailings, sludge and slime ponds associated with the mineral extraction and processing industry are good examples of such phenomena. In the case of tailing ponds associated with the mining and mineral processing industry, the interaction of surface-active soil particles produces a slime or sludge which is essentially a soil–fluid suspension (suspended solids) commonly known as a *solids suspension*. This phenomenon is represented by the interacting discrete particles shown in Figure 3.1 – except that in this case, the interparticle forces are dominantly net repulsive forces resulting thereby in maintaining a pseudo suspended state of the particles in the host fluid. They are commonly known, for example, as slime and slurry ponds (from phosphate and tin mining), sludge ponds (from processing of tar sands) and red mud ponds (from bauxite mining and processing). The feature common to all these ponds is the inability of the soil solids in the pond to sediment (settle).

Simple Stokesian models used to calculate and predict settling-time relationships for the soil solids in the various ponds obtained from process industries cannot account for the suspended nature of the soil solids. This is because the intermolecular forces dominate particle performance. Particle interactions are governed by interparticle forces. Simple and measurable indices can be used to characterize the interparticle forces. These include exchangeable sodium percentage (ESP), sodium adsorption ratio (SAR), pH, dielectric dispersion and zeta potential (ζ). The interesting and valuable lesson to be gained from conducting the zeta potential index tests is the flocculation or agglomeration of particles to form flocs under the right conditions, that is, conditions dictated by the intermolecular forces.

Figure 3.2 provides a scheme for determination of the presence of MUs and microstructures using zeta potential ζ measurements for dispersion stability of clay soils. Aggregated and flocculated particles are the basic building blocks for microstructures. The greater the aggregation and flocculation of particles, the greater is the size and proportion of microstructures in the system tested.

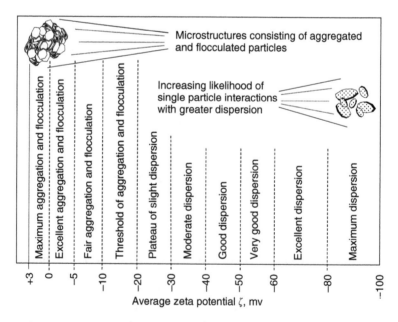

Figure 3.2 Interpretation of presence of microstructures from ranges of average zeta potential ζ measurements for dispersion stability of clay soils. Aggregation and flocculation of particles are basic building blocks for microstructures.

The energy of interaction between the particles is the result of: (a) repulsion energy and (b) attractive energy due to the London-van der Waals forces. The differences between observed interaction of particles between swelling and non-swelling soils such as smectites and kaolinites are due in a large measure to the diffuse double-layer properties of the respective soils. In swelling soils where the diffuse double-layer is well developed, interlayer swelling plays a dominant role. In kaolinite soils and other low swelling and non-swelling soils however, the dispersibility of the clay soil in the absence of interlamellar and interlayer phenomena will be a function of the interactions between the soil particles and the chemical constituents in the pore-water. Desorption or removal of adsorbed ions from the clay particles' surfaces will increase the diffuse double-layer thickness associated with the soil particles, producing thereby a pseudo particle–particle swelling phenomenon.

Desorption of Cl^- and Na^+ in a low salt content range from kaolinite particles, for example, will increase the double-layer and promote repulsion between particles. Potential-determining ions such as bicarbonates, carbonates, hydroxides etc. will also contribute significantly to the net repulsive activity between the particles. There are situations where double-layer repulsion will overwhelm van der Waals attractive force, as for example, in the case of higher proportions of monovalent cations, low salt concentration and presence of potential-determining ions.

The SAR index and per cent sodium in relation to total salt concentrations have been used extensively by many investigators in defining the dispersivity of clays. These indices take into account only the cationic species and generally ignore the anionic species. Where the alkalinity is low and the pH is neutral or near neutral, one can accept omission of the anionic species in the analyses. However, this is not acceptable when high alkalinity and high pH values are encountered. Detailed discussions of these and diffuse double-layer (DDL) models can be found in text books dealing with soil and colloid science.

3.4   Soil microstructure, macrostructure and physical integrity

3.4.1   *Influence of microstructure and macrostructure on physical integrity*

The macrostructure of clay soil is obtained from the grouping or assemblage of microstructures – as schematically shown in Figure 3.1. These microstructures will have a distribution of sizes and complement of particles. The physical integrity of a soil is determined not only by the nature, distribution of the microstructures, but also by the bonding and interaction forces between particles within the microstructures and between microstructures.

Strictly speaking, there are two groups of bonds that exist between particles in microstructures and between microstructures themselves. Both these groups are responsible for the development of the properties and characteristics of the microstructures and the macrostructure itself. The first group of forces and bonds deal directly with particle-to-particle interaction. The second group of forces and bonds are those that are developed between particles with mediating forces and bonds from the solutes in water and the water molecules themselves.

Short-range forces such as those developed as a function of ion-dipole interaction, dipole-dipole interaction and dipole-particle site interaction are of considerable importance in arrangement of particles in a microstructure. These are known as intra-microstructure and inter-microstructure bonds and interactions. Changes in soil macrostructure can occur as a result of changes in the inter- and intra-microstructure bonds and interactions.

3.4.2   Structure-forming factors

A considerable number of smectite clays have been formed in nature by weathering processes in rock but few of them are exploited commercially because of variations in composition. Those making up the major part of exploitable smectite mineral resources were formed by sedimentation and they are the ones that are the focus of this book.

The internal factors that have an impact on their microstructural evolution are interparticle forces, chemical impact by cation exchange, mineral conversion and precipitation (cementation). The external factors are temperature and strain induced by imposed stresses such as those generated by compression forces or loads.

3.4.3   Elementary constituents

By all accounts, smectitic clay deposits have their origin in slowly settling discrete particles or aggregates of particles: (a) originating as deposition of airborne ash, (b) erosion of older or ancient sediments, (c) product of turbidity currents and (d) depositions from large slope failures of sedimented clays. These different origins impact on the nature and distribution of the MUs. It is clear that homogeneity is not an expected feature of these kinds of clays.

We will deal here with sediments formed by successive, slow accumulation of small particles or aggregates. Their physical constitution depends very strongly on the electrolyte composition of the depositional fluid. The various surface-active forces associated with the particles, and their interactions with the electrolytes in the host fluid will dictate interparticle actions – and ultimately the micro and macrostructure of the sedimented soil. These have been discussed in general in Chapter 2. It is necessary to point out that the variations in size of the settling particles imply that a proper soil-structural

description requires one to define what constitutes a representative elementary volume (REV). This is best achieved by considering the REV to be a scale-dependent parameter that will be different for different purposes. It is essential to relate structural data to bulk soil properties.

3.4.4 Interparticle bonds

The interparticle bonds are:

1   Primary valence bonds between particles
2   London-van der Waals forces
3   Hydrogen bonds shared between adjacent particles
4   Bonds shared between sorbed cations on adjacent particles
5   Particle-connecting precipitates serving as cement.

Primary valence bonds

Except for cementation bonds, primary valence bonds are the strongest interparticle attraction forces: the activation energy for breakage normally exceeds 30 kcal/mole. They are dominant in heavily consolidated clays where the crystal lattices of adjacent particles are in contact. For smectites this state requires an effective pressure of about 200 MPa.

London-van der Waals forces

These forces are of importance when uncharged colloidal particles come very close. The bond strength is very low, that is, less than 1 kcal/mole.

Hydrogen bonds

Hydrogen bonds are weak (1–3 kcal/mole) but their number can be large and the net attraction force will thus be high. In principle, the coupling between adjacent smectite particles can be of the type shown in Figure 3.3.

Bonds by sorbed cations

The bonds are of Coulomb type, and the integrated attraction force between neighbouring but not contacting particles is significant. The approximate activation energy is intermediate to that of hydrogen bonds and primary valence bonds, that is, on the order of 10–15 kcal/mole. Figure 3.4 shows a schematic of this bonding mechanism.

Dipole-type attraction between particles with different charge Depending on the equilibrium or near-equilibrium pH of the immediate microenvironment, the edges of clay particles can be positively or negatively

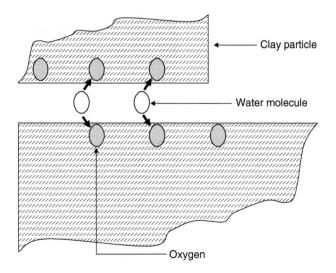

Figure 3.3 Hydrogen bonds established by water molecules located between particles (broken lines are boundaries). The arrows represent hydrogen bonds
between oxygens in the particle surfaces and the water molecules.

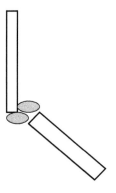

Figure 3.4 Bonds by sorbed bi- or polyvalent cations shown as ellipses.

charged. Under normal conditions the edges are positively charged and the
resulting particle coupling is of the type shown in Figure 3.5.

Bonds by organic substances   The attraction forces are primarily due to
hydrogen bonds and purely physical coupling is obtained by embracing
hyphae and flagellae. The bonds are flexible and can sustain large strain.
However, their strength is very much dependent on the environment and
can be short-lived.

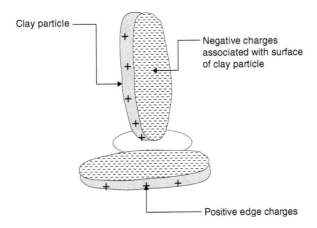

Clay particle

Negative charges
associated with surface
of clay particle

Positive edge charges

Figure 3.5 Dipole type bonding of particles with non-uniform charge distribution.

Cementation bonds

Precipitated matter binding particles together can develop bonds with strengths that can approach that of primary valence bonds (>30 kcal/mole). Their practical importance can be substantial, depending on the amount and nature of the precipitation as shown for example, by the difference between plastic clay and claystone. An example is shown in Figure 2.19.

3.4.5   The bond strength

The strengths of individual bonds cannot be readily determined by simple experimental methods. However, they can be evaluated if the number of activated bonds can be estimated. One way of doing so is to perform shear or viscometer tests of clay gels with estimated number of particles or to perform erosion tests [4]. Figure 3.6 shows the rheograms for kaolinite suspensions with 9.1% solids concentration for those suspensions established with various concentrations of NaCl, and the rheogram for a 4.85% kaolinite solids concentration in a 10 meq/L $NaHCO_3$ solution. The shear stress τ_o shown in the figure identifies the critical shear stress needed to initiate flow of the soil suspension. The soil suspension with 3000 meq/L NaCl has been used to highlight this flow initiation characteristic, and the Bingham yield stress τ_β is determined by the intercept on the abscissa of the linear portion of the flow curve.

The shear stress τ in the linear range of the rheogram can be described by the relationship:

$$\tau = \tau_\beta + \eta D$$

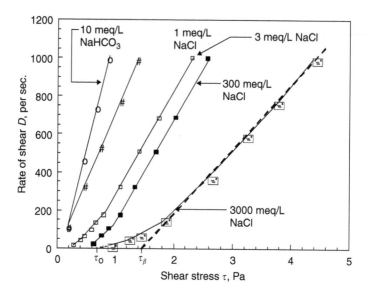

Figure 3.6 Rheograms for sodium kaolinite. Solids concentration = 9.1% w/w and 4.85% w/w for NaCl series and NaOH series respectively. Contraves Rheomat 15 rotating cylindrical viscometer with 15 different shear rates [4].

where D is the shear rate and η is the slope of the linear portion of the curve in Figure 3.6. The Bingham yield stress τ_β is a measure of the bond strength between various microstructures that comprise the total system, and the rate-dependent term ηD is the resistance offered by the interparticle forces to shear displacement. Bonding between MUs is evident at soil solutions ≥ 3 meq/L NaCl. Typical MUs are shown in the scanning electron microscopy (SEM) picture in Figure 3.7. SEM views of the overall macrostructure of the soil in suspension shows these units in edge-to-face arrangement consistent with the positive edge and negative surface charge distribution characteristics of kaolinites.

Organic bonding between particles and MUs can be observed in the rheogram for kaolinite mixed with the polysaccharide Xanthan in Figure 3.8. The results are compared with the kaolinites reported in Figure 3.6. The polysaccharide is double stranded, anionic and unbranched with dimensions equivalent to 4 μm.

Whilst the τ_0 and τ_β values for the Xanthan–kaolinite mixture shown in the diagram are lower than for the kaolinite suspension in 3000 meq/L NaCl, Yong [4] reported that with increasing proportions of Xanthan, τ_0 and τ_β will also increase. Xanthan provides a bridging bond between the positive edge charges of the kaolinite particles and microstructures. Examples of indirect means for determination of the influence of cementation and

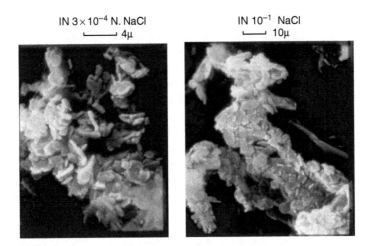

Figure 3.7 Soft kaolinite clay. Left: Relatively open, homogeneous particle arrangement in 3E–4 NaCl solution. Right: Coagulation in 0.1 N NaCl solution. The poor gel-forming potential of kaolinite causes structural collapse at dry densities lower than about 1000 kg/m³ and sedimentation of large aggregates [4].

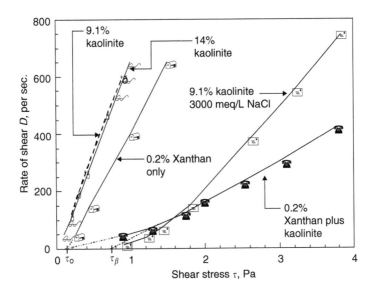

Figure 3.8 Rheograms for kaolinite mixed with the polysaccharide Xanthan [4].

interaction bonds between microstructures on the integrity of clays can be found in the Champlain Sea clays. When these clays are disturbed and remoulded, they tend to flow as a very thick viscous fluid.

Results of triaxial and other physical tests on these clays have shown that their sensitivity and brittleness are the result of cementation bonds between particles and MUs, and that the arrangement of the microstructures is in an open flocculated form [5–7]. Studies conducted to determine the bonding forces and mechanisms responsible for the extreme sensitivity of these marine illitic-chloritic clays and other Scandinavian marine clays have included leaching of salts and cementing agents, and addition of chemicals to determine changes in remoulded strengths [8]. Evaluation of the porewater chemistry and amorphous materials and their relation to interactions and bonding has been made by Moum and many others [9–13].

Tests on mechanical properties such as compressive strength show that contributions from the microstructure bonds and interactions are measurable and can be significant at confining pressures below the preconsolidation pressure. Figure 3.9 is an indication of such a phenomenon. The hump shown in the diagram describes the contributions from the microstructural bonds. In contrast, a non-structured soil will generally obey linearity in performance when the information on the strength tests is expressed in the manner shown in the figure. Remoulding of the natural soil prior to strength testing will also show linearity in performance.

The reported studies on leaching with salts and other chemicals also indicate how bonding between the microstructures can be sensed. Removal of cementatious materials between particles obtained by selective dissolution

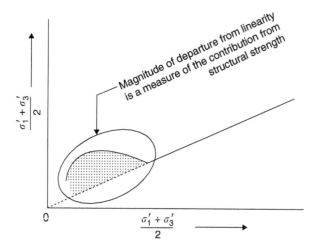

Figure 3.9 Contribution of microstructure bonds and cementation to triaxial compressive strength. σ_1' and σ_3' refer to effective major and minor principal stresses, respectively.

with various kinds of reagents (e.g. ethylenediaminetetraacetic acid (EDTA)) results in dramatic decreases in the size of the hump shown in Figure 3.9.

Results from leaching of soils with salts (salt solutions) show that interactions between particles and microstructures can be increased or decreased depending on the nature and concentration of the salts used – the primary interactions affected being those obtained in the diffuse double-layers. Depending on the initial bonding of the microstructures, the bond strengths may or may not be dramatically altered with salt leaching. This depends on the ability of the changed interactions to counter the bonded nature of the material by increased repulsion forces (swelling) between particles and microstructures. If leaching decreases the swelling potential, the result of leaching with salts will see a contribution to the initial bonded strength of the soil.

Another useful procedure for deducing the presence and influence of microstructures can be obtained by introducing bonding material into artificially prepared clays such as laboratory-prepared kaolinite soils. The results shown in Figure 3.10 provide test information on fluid transport in relation to clay soil composition as deduced from a column leaching test. The addition of 10% amorphous silica to the kaolinite essentially produces the same kinds of results on microstructure development as obtained with a combined amorphous silica–carbonate system. This is due to the fact that with the carbonates, cation bridging and hydrogen bonding would develop

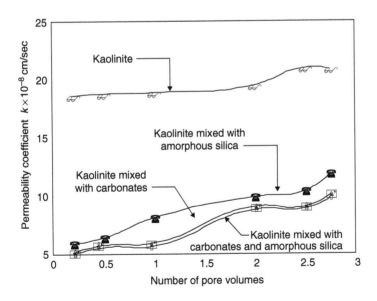

Figure 3.10 Variation of calculated hydraulic conductivity coefficient (k in the figure) with number of pore volumes discharged in fluid leaching experiments using data from Darban [14].

polymerization of the clay complex. When there is only addition of amorphous silica, the amorphous coatings of the kaolinite particles (resulting from electrostatic attraction between the clay particles and the amorphous silica) will cause the clay particles surfaces to assume the characteristics of the amorphous silica.

The three mechanisms for development of microstructure because of the interaction between the amorphous silica and the clay minerals include: (a) electrostatic attraction between the clay particles and amorphous silica, (b) cation bridging through adsorbed cations and (c) hydrogen bonding between the kaolinite minerals and the amorphous silica.

The data obtained by Darban [14] for partitioning experiments with the same kaolinite mixed with 10% amorphous silica, using lead nitrate leachate at pH 3, shows that the effect microstructure can be deduced from the amounts of Pb sorbed or partitioned onto the soil solids (Figure 3.11).

Although relatively little difference in total retention of Pb in the kaolinite and kaolinite–amorphous silica soils is obtained, there is a significant difference in the sorption (partitioning) of Pb onto the soil solids of the two soil systems. Examination of the sorption of Pb by the soil solids of the K and KS soils shows a distinct difference between the two – with a significantly higher sorption or partitioning of Pb by the KS soil solids. The remaining Pb concentration in the pore fluid, which is obviously higher for the K soil as opposed to the KS soil, suggests that a polymeric microstructure exists for the KS soil. The MUs appear as packets and flocs, and may

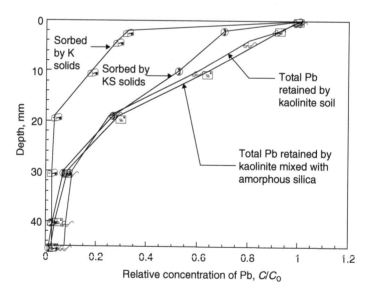

Figure 3.11 Partitioning of Pb onto soil solids in column leaching experiments using a lead nitrate leachate. 'K solids' and 'KS solids' represent kaolinite and kaolinite–amorphous silica soil solids respectively [14].

or may not exhibit some measure of polymerization depending on whether other soil fractions are available to establish bridging bonds.

3.4.6   Impact of porewater chemistry on the arrangement of clay crystallites in the settling sediment

In the sedimentation process, the forces of interaction between settling particles causes aggregation, as is well known from colloid chemistry – and as illustrated by the experiments with kaolinite referred to in the preceding section. In the slow formation of deep-sea sediments, the coupling of particles will result in the establishment of clay gels with a mechanical strength that is just sufficient to carry the successively increasing overburden. In essence, an equilibrium state is achieved between the interparticle forces and the overburden pressure or load. The fundamental principle that soils become normally consolidated if deposited under water applies to any sediment, provided that no chemical processes such as dissolution and cementation occur.

The degree of homogeneity of the initially formed clay gels depends on particle charge and size and the hydration properties of the particles. The difference between smectites and illites, for instance, is striking because the smectite particles are smaller and tend to repel each other through interacting DDL (i.e. repulsion due to DDL mechanisms) at the basal surfaces as discussed in Chapter 2. With slow sedimentation, the clay matrix has the opportunity to become very homogeneous. The slower the sedimentation rate, the more homogeneous is the clay matrix. As will be discussed in Chapter 4 the microstructural constitution of natural smectite sediments is also more homogeneous than similar clays prepared by drying, grinding and compaction after ultimate water saturation.

A very useful technique to obtain parallel-oriented arrangement of montmorillonite particles is to sediment very dilute montmorillonite suspensions – about 0.1% solids concentration or less. By using small quantities of the suspension in appropriately shallow settling dishes, one can obtain sediments that, when subject to slow one-dimensional dehydration, will yield thin wafers of montmorillonite. The process is admittedly slow. However, when a number of these dry wafers are stacked one on top of the other and allowed to uptake water under a controlled load, one would obtain a compact sample with an almost perfect parallel orientation of discrete particles. This technique has been used by researchers in studies to examine the applicability of the various DDL models, and to better explain the interlamellar and intralamellar behaviour of smectites.

The effect of porewater chemistry in the formation of natural sediments is obvious: if the salinity is high, the system of settling particles will flocculate, resulting in the formation of large and dense aggregation of particles. On the other hand, if the salinity is less high, the aggregation of particles will be small and the resultant gel obtained with the smaller aggregate units will be more homogeneous and will contain smaller voids (Figure 3.12). We will use the term *aggregates* to mean an aggregated group (aggregation) of particles. Figure 3.13 shows discrete aggregates in low-electrolyte river water. Their sizes typically range between about one micrometre and many hundred micrometres.

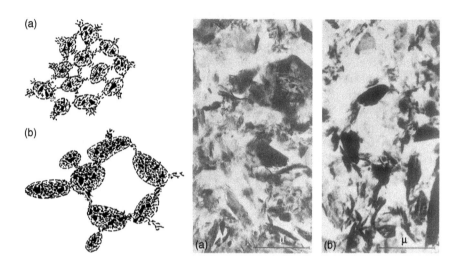

Figure 3.12 Clay aggregates in soft illitic sediments. Left: schematic drawings of (a) fresh-water sediment with small soft aggregates separated by small voids. (b) Marine clay sediment with the same bulk density having large and dense aggregates separated by large voids. Right: TEM pictures of natural clay, mostly illite and chlorite particles, in fresh (left) and salt water (right), respectively [15].

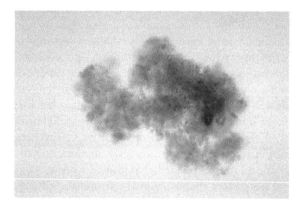

Figure 3.13 Micrograph of 20 μm aggregate of illite/mixed-layer particles in low-electrolyte river water (Elbe) photographed by optical microscopy (Photo: Kasbohm) (see Colour Plate VIII).

A clay that is sedimented under marine conditions and later percolated by fresh water will suffer an alteration in its structure. The aggregates of particles (MUs) will soften and a drop in average shear strength will result – as explained earlier. Remoulding of soft sediments of this type will result in a complete dispersion of the particles, turning it into 'quick clay' [14, 16]. The phenomenon is particularly obvious for clays with poor gel-formation potential such as illite and kaolinite, but less obvious for smectite. This is because of the capability of the smectite to re-establish interparticle bonds and also because of its thixotropic nature.

Soft smectite clay with sodium ions in the exchange positions undergoes partial interlamellar dehydration if the porewater becomes rich in calcium. Partial dehydration occurs in the interlamellar spaces (interlayer spaces) resulting in the development of larger void spaces between the particles (i.e. particles composed of stacked lamellae). The resultant effect is an increase in the hydraulic conductivity of the initially soft smectite.

For the dense smectite clay, on the other hand, since the clay is already populated by a limited number of hydrates in its original sodium state, further contraction of the lamellar stacks will not easily occur. This issue is discussed in detail in Chapters 4 and 5.

The nature and strength of the interparticle force field depends on the porewater chemistry. This holds for all types of smectites, mixed-layer clays containing smectite and illite. However, it is the differences in net particle charge that makes the impact of porewater chemistry much stronger for smectites than for mixed-layer minerals and illite.

3.4.7 The bond strength spectrum – the basis of stochastic mechanics

The varied nature of interparticle bonds and their respective strengths, combined with the heterogeneity of the network of clay particles and the MUs (i.e. heterogeneity of the macrostructure), mean that the reaction of an element of clay to imposed stress will be complex. As with any mechanical or physical structure, the integrity of the structure is compromised by the *weakest link*. In the case of the clay soil, this will be the weakest bonds between particles and between MUs. Under stress, the weakest hydrogen bonds will fail first, thereby transferring to stronger junctions established by bridging cations and dipole-type attraction, and further to cementations bonds and primary valence bonds in mineral/mineral contacts. The progressive nature of varied bond breakage and collapse can occur over a long time period – obviously dependent on the original types and distribution of bonds, and on the nature of the imposed stresses. Figure 3.14 shows a clay specimen under load. A microscopic portion of a volume element is shown at the top right of the schematic. Application of a load imposes stresses on the total structure of the clay soil, and in turn on the various elements (particles and MUs) of the total soil structure. The sequence of bond breakage

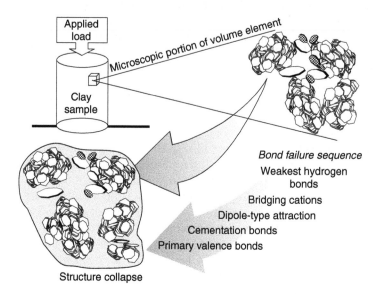

Figure 3.14 Schematic showing failure of the total structure of a clay sample under load because of collapse of the various elements (particle shear displacement and microstructure collapse) of the total structure. The sequence of bond collapse and the progressive nature of the bond failure result in structure failure occurring over a protracted time period.

and the collapse of the total structure is the accumulation of the collapse of MUs. This is observed as deformation of the sample. The progressive nature of the collapse of the various elements is responsible for the creep rate of the sample not being proportional to the applied shear stress.

This will also further explain why creep can lead to failure long after the application of a critically high shear stress. The first-mentioned effect is due to the fact that strain (deformation of the sample) occurs because the bonds with lower energy will be exhausted – resulting thereby in the participation of stronger bonds with higher energy levels. The second effect is related to a total loss of low energy bonds at a critical strain, and to an ultimate stage where one encounters an insufficiency of bonds of high energies [17]. We will return to this matter in subsequent chapters.

3.4.8 Compression

Compression of soils in nature with porewater drainage is generally known as *consolidation*. This conforms to soil mechanics terminology where the process of consolidation is defined as compression (of a soil) in conjunction with porewater extrusion during the compression process. The consolidated state of a soil is reached when stress equilibrium is obtained between the various elements (particles and microstructural units) that constitute the

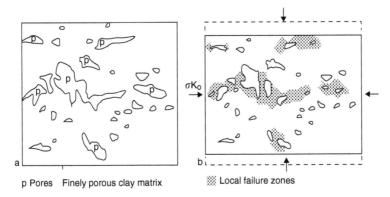

p Pores   Finely porous clay matrix   ▦ Local failure zones

Figure 3.15 Microstructural changes by compression. The largest voids become
compressed first in conjunction with shear-induced breakdown of the
clay matrix adjacent to them [18].

total soil structure. This means that the mechanical strengths of the particle
and microstructural network that oppose the effective stress imposed by
the overburden are at equilibrium. By and large, an increase in the effective
stress will result in a pore space reduction in accord with a power law. For
an effective overburden pressure of 50 kPa corresponding to a depth of a few
metres, the porosity of most clays is about 70%, with a dry density of about
800 kg/m^3. For an effective overburden pressure of 30 MPa, the porosity is
around 20% with a dry density of about 2200 kg/m^3, corresponding to a
sample location about two kilometres below the sea bottom. At this level or
at this porosity and dry density, the compact nature of the soil is such that
there are no continuous void systems. Transport of fluids and dissolved
chemical species will be by diffusion mechanisms.

The mechanisms involved in compression are illustrated in Figure 3.15. The
schematic shows the microstructural strain that a clay undergoes at loading
[18]. The most significant change that is observed is the compression of the
larger voids under increased pressure in conjunction with the breakdown of
the clay particle network – primarily by shearing. The consolidation rate is
determined by the dissipative characteristics of the porewater in the stress zone,
and by the rheological behaviour of the undisturbed part of the clay matrix.
The rheological properties of the soil will totally control the deformation rate
of the soil when excess porewater no longer exists. In deep sediments, total
dissipation of excess porewater pressure can take many thousand years.

3.4.9   Shearing

The impact of high shear stresses on natural clay microstructure in slopes
or under the foundation of buildings has been the subject of soil mechanics
research for decades. Overstressing causes breakage of interparticle bonds

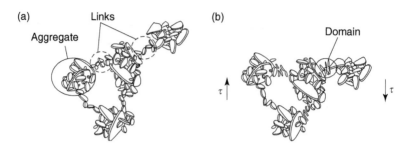

Figure 3.16 Failure in clay at critical stress conditions. (a) undisturbed state. (b) Partial destruction by shear failure of weak links that turn into local groups of aligned particles ("domains").

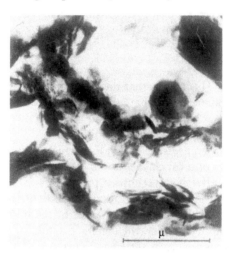

Figure 3.17 TEM micrograph of microstructural changes in soft illitic clay by particle alignment in the shear plane at triaxial testing to failure.

and alignment of particles particularly in the weak parts of the microstructure – as indicated in Figure 3.16. Investigations of specimens extracted from triaxial tests in which the cell pressure, that is, the minor principal stress, was kept constant and the axial, major principle stress was increased stepwise until failure, have shown very obvious alignment of particles in the macroscopic shear plane (Figure 3.17) but significantly less structural changes only millimetres away from the shear plane.

3.4.10   Creep

The results from a large number of microscopy studies have contributed to the present understanding of how shearing affects clay particle networks. These have formed the basis for theories to explain processes of creep and consolidation on the microstructural level. The most important practical examples in nature are slope failures in thick soft clay layers. These occur as a result of particle bond breakage and alignment, leading to a loss in shear strength, and consolidation and densification under conditions of constant effective pressure. This will be discussed further in subsequent chapters.

The complex microstructure characteristic of soil in general and clays in particular shows the need to take into consideration heterogeneity in models of plasticity and creep. One needs to take into account the activation energy barriers of the various or different types of MUs in construction of the analytical model. This requires knowledge of the various types of interparticle bonds involved in development of aggregates of particles – as discussed earlier in this chapter.

3.4.11   *The physical state of porewater in smectite clay*

The physical interaction of porewater and the mineral phase is different for different soils. In practice, the porewater is generally considered to be a separate phase in soils, with the properties of bulk water. For smectites, and to some smaller extent for illite and chlorite, although the water in the system is classified as porewater, it does not have all the properties of bulk water. As implied by the molecular dynamic theory referred to in Chapter 2, the viscosity of the adsorbed water is higher than that of free water. However, the mobility of the molecules is higher than in the interlamellar space. It is assumed that no more than three hydrate layers are adsorbed on the free surfaces and that only the first two have physical properties that deviate significantly from those of free water. In montmorillonite, a certain fraction of the porewater is in interlamellar space (micropores), and the rest in the void spaces identified as *macropores*. Using relevant particle size distributions and microstructural models, one can obtain the weight fraction of interlamellar water for Na and Ca in interlamellar positions in Figure 3.18.

For low bulk densities (of the soil), since the water content is low, the proportion of interlamellar water to the bulk water in the macropores will be high. Using ω_i to denote the quantity of interlamellar water and ω_b to denote the quantity of bulk water in the macropores, the ratio of ω_i to $\omega_b = \Omega$ will be high. For the high bulk density situation, although the total water content is higher, we need to note that there will still only be a maximum number of hydrate layers in the interlamellar space. Accordingly ω_i will likely be the same value as for the low bulk density case. This means that the proportion of the interlamellar water to bulk water in the macropores Ω, will be lower for the high bulk density situation since the ω_b for this case will be considerably larger than for the low bulk density case.

For unsaturated bulk densities lower than about 1500 kg/m^3, about 40% is bulk water or water in the macropores. This means that the DDL forces in interlamellar space and between contiguous particles control the rheological performance of smectites such as coagulation and thixotropy.

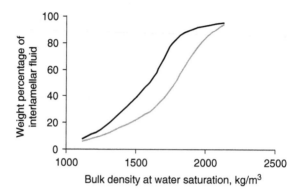

Figure 3.18 Theoretical relationship between bulk dry density in kg/m³ and the content of interlamellar water expressed in percent of the total porewater content for montmorillonite-rich clay [1, 16]. The upper curve represents Na and the lower Ca clay.

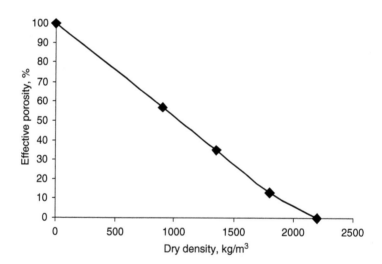

Figure 3.19 Effective porosity, that is, the volume fraction of montmorillonite-rich clay (MX-80 bentonite) that hosts free water.

We see from Figure 3.18 that the type of adsorbed cation will affect the uptake of water into the interlamellar space – that is, we can obtain a different number of hydration layers (hydrates) in interlamellar space. The effective porosity, that is, the volume fraction of voids between the stacks of lamellae of the total volume, can be calculated from these data and is given in Figure 3.19 as a function of the dry density.

For higher bulk densities at fluid saturations greater than about 1900 kg/m³, the major proportion of the water is in the regions that are not strongly affected by DDL forces. This explains why even very significant changes in porewater electrolytes do not have any major effect on either the hydraulic conductivity or the swelling pressure. For other smectite species that can hold only two interlamellar hydrates, the behaviour is the same as for montmorillonite with Ca as the major adsorbed cation.

3.4.12   *Impact of the arrangement of smectite crystallites on the physical properties of interstitial water*

The organization of the stacks of lamellae affects the physical state of the porewater. Thus, the water in micropores is affected by the confining mineral surfaces as concluded from (proton) nuclear magnetic resonance tests [19]. Such studies have led to the conclusion that the viscosity of water in pore spaces or voids smaller than about 5 nm (50 Å) is higher than in free water. Some reduction in fluidity may exist also in voids as wide as 10^{-2} μm, corresponding to the narrow wedge-shaped space between adjacent solid mineral particles. Table 3.1 is a proposed scheme for categorizing voids, with estimated impact on the viscosity of the porewater. It suggests deviation from Darcy's law for high densities and complete flow tightness for a dry density of 2200 kg/m3 (2390 kg/m³ at complete water saturation).

3.4.13   *Temperature impact*

Sedimentation and formation of clay particle networks are not significantly affected by water temperatures below a few 'tens of centigrade' (about 40°C). Whilst the sedimentation rate increases with the drop in viscosity that is caused by heating, the counteracting increase in Brownian motion of colloid-sized particles and water convection is of greater importance for evolution of the microstructure.

Table 3.1 Void classification

Void category	Size range, μm	Impact on porewater viscosity
Coarse	> 200	None
Medium	20–200	None
Fine	2–20	None
Micro	0.2–2	None
Ultra	0.02–0.2	Negligible
Nano	0.002–0.02	Very slight increase
Pico	0.0002–0.002	Significant increase
Sub-pico	<0.0002	Very significant increase

Heating of a natural clay layer increases the porewater pressure and reduces the effective pressure and shear strength. The combination of these can cause stability problems for repositories constructed in deep argillaceous rock. In shallow soil where desiccation/wetting can occur, the microstructure undergoes very significant changes. This situation usually leads to aggregation, fissuring and strengthening of the matrix between fissures. These changes can be accentuated by precipitation of chemical elements such as iron, sodium and calcium-forming oxyhydroxides, chlorides and carbonates leading to formation of an 'overconsolidated' dry crust.

3.4.14   Mineral stability

Chemical changes of the minerals in a sediment start immediately after their deposition. This can be clearly seen in the isothermal corrosion of primary mineral particles and precipitation of neoformed substances in recent clays. Ancient sediments have often undergone significant changes by dissolution. This is a subject that has been studied by the geochemical scientists, and will be further treated in Chapter 7.

In nature, deep smectite deposits show considerable persistence and stability. There are deposits that have existed for many millions of years at depths down to a couple of kilometres at estimated effective pressures of at least 30 MPa and temperatures up to about 60°C [2]. Deeper down, as shown by the sediments in the Atlantic Ocean and the Mexican Gulf, there is a systematic change of these deposits to illite and chlorite [20, 21].

The major mineral transformation mechanism is conversion of smectite to mixed-layer minerals or a separate illite phase by dissolution/precipitation. The rate of conversion is controlled by accessibility to potassium. These processes involve contraction of the smectite stacks of lamellae and associated growth of the voids. The water in the voids may become pressurized, leading thereby to an additional consolidation process and higher density under the overburden pressure. Similar processes may take place in smectitic engineered barriers as described in Chapters 5 and 7. These processes affect the isolation potential of smectite barriers for waste isolation by increasing their hydraulic conductivity and reducing their expandability and hence their self-sealing ability. This matter is of fundamental importance.

3.5   Methods for microstructural analysis

3.5.1   Terminology

The arrangement of soil constituents (soil fractions) has been historically termed *soil structure*. The general term is used here with the prefix *micro* for features that cannot be discerned with the naked eye and *macro* when this is possible. Studies on identification and characterization of soil structure have been performed in many scientific and technical disciplines such

as sedimentology, geology, soil physics, soil mechanics, agriculture and forestry. The earliest systematic studies concerned coarser soils and were conducted in the field of sedimentology in conjunction with development of fundamental physical laws for hydraulic conductivity and erosion. Investigation of clay microstructure was not comprehensive until advanced microscopy methods were developed.

3.5.2   Techniques

One of the major problems in the study and identification of microstructural features of a soil is preparation of a representative sample of the soil that allows the natural particle arrangement to be preserved and visualized. Work in this area started in medicine and agriculture a few hundred years ago, and is still ongoing globally – with a view to improving the techniques.

For traditional methods, that is, microscopy using visible light or electron radiation, thin sections must be prepared to permit observation of microstructural details. For microscopy using on-falling radiation it is sufficient to prepare an even surface without damaging the microstructure.

Transmissive radiation, the impact of section thickness

The thickness of the section from which micrograph pictures are taken is of fundamental importance – as shown by Figure 3.20. The ideal thin section is a two-dimensional plane. The thicker the section, the smaller is the possibility to identify voids included in the clay matrix. The question of what thickness one can accept depends on the required resolution and the need for discerning microstructural details.

Specimens are prepared by impregnation of the sample so that the porewater is replaced by a suitable substance that hardens with minimum impact (disturbance) on the microstructure. Preparation of sections for light microscopy with a thickness of 10–30 μm is made by ordinary microtomes equipped with special steel or diamond knives. Light microscopy using

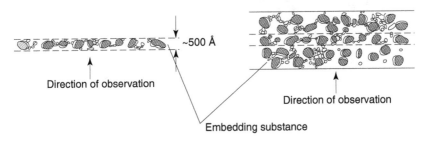

Figure 3.20 Influence of section thickness on the interpretation of microstructural details.

special wave-lengths and optics can be used for identifying microstructural details with a size down to about 3 μm.

For ordinary transmission electron microscopy (TEM), much thinner sections must be prepared (i.e. 20–100 nm), for which ultramicrotomes have to be used. Details with a size of down to 1 nm can be discerned. This makes this method superior to all other techniques. Cutting of frozen samples by use of cryotomes has been tried. However, the freezing process causes some disturbance and hence renders this technique questionable. Neutron bombardment for preparation of thin sections has been tried. It is however very difficult and unreliable.

In the late 1960s and 1990s, comprehensive work was directed towards investigating 10–20 μm thick sections of clay with its natural water content, using high-voltage electron microscopy. These studies confirmed that TEM images of clay impregnated with certain substances are representative.

The ideal section is a two-dimensional lamella. The thicker the section the smaller the possibility of identifying voids or other structural features included in the clay matrix. However, the extremely thin sections obtained by 'ultramicotomes' become very fragile and experience shows that 20–30 nm is a practical minimum for preparing sections of polymer-impregnated clay. At this thickness however, we encounter problems with folding, curling and breakage of the thin sections. It is advisable to inspect the section with an ocular device to identify useful sections (Figure 3.21) [22].

On-falling radiation

Preparation of representative surfaces of undisturbed clay structure for investigation and examination by optical and scanning microscopy using on-falling radiation is difficult. These techniques are primarily used for: (a) general characterization of fine soil, (b) identifying fossils in paleontology and (c) identifying special minerals in mineralogically oriented projects. However, some use of this type of microscopy has been made for

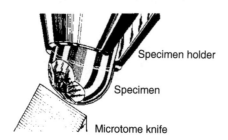

Figure 3.21 Close-up view of knife and specimen in ultramicrotomy. The cutting process and picking up of thin sections from the liquid in the trough are observed through a strong lens and intact sections saved for microscopy.

microstructural characterization. In the method for peeling that is necessary to allow one to reach down below a fracture surface into the undisturbed clay matrix, a suitable peeling tape is repeatedly applied and removed. The process is terminated when a surface suitable for examination is obtained [23]. Although the technique may be simple, the rough or rugged surfaces obtained makes for an uncertain definition of voids and aggregates. It is not unusual, for example, for the ripped-off fragments to leave behind depressions that may not represent true voids. Further artifacts arising from the successive peelings can also be cited. Damage to the microstructure cannot be avoided.

Present state-of-the-art

At present it is believed that impregnation of clay samples with a suitable well-hardening substance for replacing the porewater is the best method available for sample preparation. However, much work is ongoing and expected to lead to equally good or better techniques. A promising method is the application of tomography that is specially adapted to soil structure analysis. The present resolution power of about 10 μm is sufficient for some general characterization of coarser soil, but not for detailed analysis of smectite clays [24]. The method, which can be refined, makes use of commercial microfocus X-ray computerized tomography equipment. The images are analysed by use of computer software. From such analyses, it is possible to evaluate: (a) changes in structure during sample hydration or dehydration and (b) the tortuosity factor of diffusion paths.

3.5.3 *Methods for impregnation of clay specimens*

A large number of impregnation techniques have been used to stabilize the particle network of clays. The problem that requires attention is the replacement of the porewater by a monomer without disturbance of the network during the replacement phase or subsequently during polymerization of the monomer.

   A common procedure is to contact the sample, which must be confined in a small cell equipped with filters, with ethylene alcohol of successively increasing concentration. This is followed by replacement of the alcohol with a suitable monomer such as butyl/methyl methacrylate, Vestopal, SPURR, LR White Hard Resin. These are well-known commercially available materials. After polymerization at 50–60°C for some tens of hours the sample becomes hard enough to be sectioned by a microtome [22, 25]. For smectites, alcohol treatment is questionable since it may cause irreversible microstructural changes, especially for soft clays. For such cases, water-soluble monomers such as Durcupan appear to be preferable. Other ways of impregnating such clays is to use high-polymer versions of viscous Carbowax at increased temperatures, and to cut or grind thin sections of the clay after cooling to very low temperatures.

The 'traditional' acrylate method

The procedure for acrylate preparation is as follows [22]:

1   A prismatic specimen with a base area of a few square millimetres and a length of about 10 mm is cut from the water-saturated sample such that its orientation *in situ* can be defined.
2   Place the specimen in a solution consisting of 50% by weight of ethyl alcohol and 50% distilled water for 30 minutes, followed by placement in 90/10 ethyl alcohol/water and 99.5% ethyl alcohol for 5 minutes each.
3   Following step 2, transfer specimen to a solution consisting of 85% by weight of butyl methacrylate and 15% methyl methacrylate for 45 minutes, the process being repeated twice.
4   Finally, transfer the specimen to a solution of 98% monomer and 2% 2,4-dichlorbenzoylperoxide (EMW) catalyst for 90 minutes. Polymerization is obtained by heating to 60°C for 15 hours.

For impregnation of expandable minerals such as smectites, the sample must be confined in a strong filter-equipped cell. This procedure is essential to restrict the sample from swelling during the impregnation process. Otherwise, the original microstructure will not be preserved [25].

'Direct' methods

Impregnation of a smectitic clay without prior drying can be made by repeated impregnation of special water-soluble substances such as Durcupan and Carbowax. This procedure removes the porewater and replaces it with a substance that becomes sufficiently hard for microtomy.

It is also possible to obtain representative micrographs by gentle drying of smectite clays and subsequent impregnation with butyl/methyl methacrylate with catalyst. This treatment has been found to yield the same swelling pressure as the original clay with its natural water content, indicating thereby that the microstructural constitution is similar for both states (Figure 3.22) [25].

The influence of drying prior to impregnation needs serious consideration. Dehydration by heating clay specimens at 105°C or by freeze-drying causes shrinkage of the stacks of flakes as a result of complete loss of interlamellar water. The latter process causes less overall shrinkage and hence partial preservation of the continuity of the particle network. The difference is small for dense smectites but can be considerable for soft ones. Slow drying at a relative humidity (RH) of 40–50% followed by quick vacuum treatment is estimated to give slightly less or the same disturbance as freeze-drying – at least for high densities (Figure 3.23). This method is promising as a first step in impregnation of expandable clays with a density at saturation of at least 1600–1700 kg/m^3.

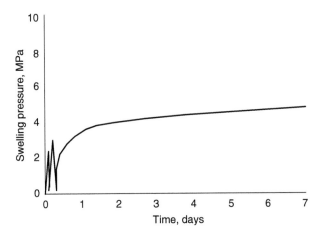

Figure 3.22 Swelling pressure built up during saturation of an air-dry MX-80 bentonite sample with butyl/methyl methacrylate. Dry density is about 1700 kg/m³.

Figure 3.23 Structural effect on smectite clay due to drying [25]. Left: Expanded hydrated stacks of lamellae. Right: Collapsed, dry stacks.

3.5.4   *Microstructural parameters*

Descriptive features

Micrographs of very thin sections obtained by techniques using light and TEMs represent almost two-dimensional sections. These can be used to determine microstructural parameters that are related to the physical properties of the clay in bulk, provided that scale effects are taken into account. This requires that the size of the section to be examined must represent a relevant or REV that increases with the grain size range [26]. For very fine-grained clays the corresponding section size can be as low as a few micrometres while for boulder clay it can be several square decimetres.

The features that can be identified and quantified are:

- Lamination, thickness and frequency of lamellae of different mineralogical components
- Discrete particles, size and shape

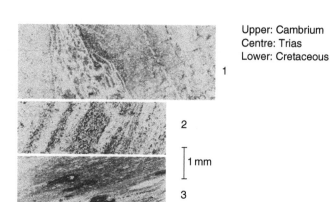

Upper: Cambrium
Centre: Trias
Lower: Cretaceous

1

2

1 mm

3

Figure 3.24 Light-micrographs of natural clay-rich materials. They all have
significant amounts of smectite.

- Aggregates of particles, size, shape and density
- Voids, size and shape
- Orientation of discrete particles, aggregates and voids.

Figure 3.24 shows micrographs taken by light microscopy of Cambrian,
Triassic and Cretaceous clay, which all have appreciable amounts of the
smectite mineral montmorillonite. The historical overburden pressure is
estimated to have been several MPa. The micrographs demonstrate the
obvious fact that primary bentonites, that is, autochthonous sediments of
volcanic origin, and secondary bentonites, which are eroded and rede-
posited – allochthonous sediments – are characterized by a rather heterogeneous microstructural constitution. If the overburden pressure had been
low, the clay matrix would be relatively open, with voids separating aggregates of different size and density. Focusing on finer structural details of
the Triassic clay as shown in Figure 3.25, one recognizes the obvious orientation seen in Figure 3.24 but no non-clay laminae is seen, meaning that
the REV for this clay is about one cubic millimetre. A special case of
allochthonous smectitic clay is moraine, which is of interest because it is
ideally suited for tight and low-compressible backfills in repositories
(Figure 3.26).

Figure 3.27 is the macroscopic equivalent of micrographs taken by scanning
microscopy, and demonstrates the difficulty in defining structural features
with respect to size and orientation. Despite this disadvantage scanning
micrographs such as this one are of value for general microstructural
characterization especially if the microscope is equipped with facilities for
element analysis (EDX).

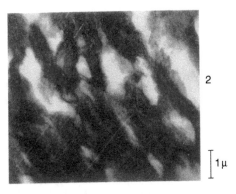

2

1μ

Figure 3.25 TEM micrograph of ultrathin section of smectitic Triassic clay from Vallåkra, Sweden.

Grain size distribution, mm

Fraction, mm	Weight percentage
<20	100
<2	90
<0.2	70
<0.02	50
<0.002	25

Figure 3.26 Smectitic moraine clay from Hyllie, Sweden.

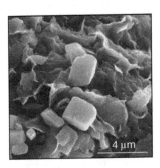

4 μm

Figure 3.27 Scanning micrograph of fractured freeze-dried smectite clay sample. Grain characteristicis can be derived from this type of micrograph. NaCl crystals are seen embedded in the clay matrix.

*Quantification of clay microstructure by
use of TEM pictures*

Quantitative microstructural data are those that can be used for explaining bulk behaviour of the soil in question and for correlating its constitution with its physical properties. In this regard, the arrangement and orientation of differently sized particles and aggregates of particles as well as of the system of voids are essential. They can best be evaluated from micrographs of very thin sections such as those used in association with TEM. An example of this can be seen in the micrograph of impregnated and ultramicrotomy-cut smectitic clay shown in Figure 3.28. The dark parts represent the most electron-sorbing objects, that is, dense rock-forming minerals, and the brightest portions represent voids filled with the impregnation resin. The various degrees of grayness are clay gels of different density [25]. A number of features can be seen. These include: (a) a certain degree of particle orientation related to the natural orientation, and (b) a weak layering tendency of the clay mass. The latter is characteristic of the German Friedland clay of Tertiary age. It has a content of minus 2 μm particles of 90% and a

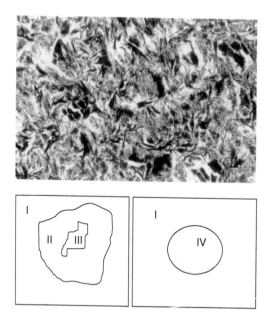

Figure 3.28 Upper: TEM micrograph of 500 Å ultrathin section of clay with about 50% expandable minerals and 1900 kg/m³ bulk density at water saturation. Darkest parts are minerals, while white parts are open voids. The bar is 1 μm long. Lower left: soft matrix (I to III). Lower right: channel section [26].

content of swelling clays (mixed-layer montmorillonite/muscovite and montmorillonite as a separate phase) of 50%.

For clay-rich soils, experience has shown that it is more useful to focus on the fraction of a section that is represented by voids since this represents the permeable part of the clay. As indicated by Figure 3.28 one can identify dense and soft parts of the particle network as well as voids that are completely open. Their respective fractions of the analysed part of the section can be determined with the aid of automatic scanning of digitalized versions of the micrographs. This makes it possible to distinguish between dense and impermeable parts of the clay, medium dense low-permeable parts and open very permeable parts. From the viewpoint of hydraulic conductivity one can generalize the microstructural pattern to consist of low-permeable and high-permeable 'channels' as in the figure.

Useful microstructural parameters introduced some decades ago were the size (maximum diameter) of discernible voids, and the P/T ratio with P representing the sectioned voids, that is, IV in the right picture in Figure 3.28, and T = the total section area. The latter can be correlated to the bulk hydraulic conductivity of natural illitic clays deposited in fresh-water and salt water as discussed in Chapter 5. Statistical treatment can be applied to suitably defined microstructural parameters, taking the solid clay matrix to consist of two major components [26], that is, (1) stacks, stack aggregates and non-smectite grains (a) and (2) gel-filled voids and unfilled voids (b). This distinction is made on the ground that the first-mentioned component is completely or largely impermeable whereas the latter offers little or no flow resistance. A further reason is that component a is largely responsible for the swelling pressure. Ion migration in this component takes place by both pore and surface diffusion. On the other hand, ion migration in b-space is almost entirely by pore diffusion. The two microstructural components are related through the coefficients F_2 for 2D and F_3 for 3D conditions, and their ratio depends on the average and individual bulk densities as defined in Figure 3.29 and illustrated in Figure 3.30 for Wyoming bentonite. This procedure for density-related microstructural characterization can be applied to any soil analysed with respect to the microstructural constitution using micrographs of very thin sections.

F_2 and F_3 can be evaluated from digitalized TEM micrographs with different degrees of grayness representing different densities. They can be converted into different colours for easy interpretation and representation of the variation of density. In the current R&D work this has been made by transforming scanned micrographs to digitalized form using the OFOTO 2 code, with subsequent colouring using the GRAPHIC CONVERTER 2.9.1 code on a MacIntosh Power PC6100/66. There are many other ways of producing and evaluating TEM pictures and improved techniques are under way.

An example of a processed picture is shown in Figure 3.31. Using only four colours, clear distinctions can be made of parts representing different

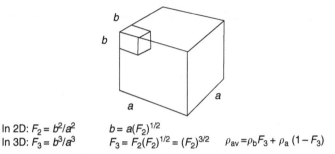

In 2D: $F_2 = b^2/a^2$ $b = a(F_2)^{1/2}$
In 3D: $F_3 = b^3/a^3$ $F_3 = F_2(F_2)^{1/2} = (F_2)^{3/2}$ $\rho_{av} = \rho_b F_3 + \rho_a (1 - F_3)$

Figure 3.29 Method for defining microstructural features in 2D and 3D. ρ_{av} is the average 'bulk' density of the clay and ρ_a and ρ_b the average density of components *a* (stacks, stack aggregates and non-smectite minerals) and *b* (soft gel fillings and open space).

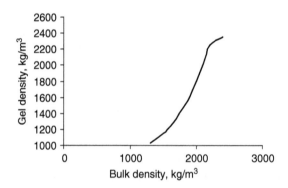

Figure 3.30 Correlation of the average gel density (ρ_b) and the average bulk density (ρ_{av}) for Wyoming bentonite (MX-80).

densities: black parts, that is, the most electron-absorbing components being the densest parts of component *a*, and red parts relatively dense parts of the same component. Green parts are taken as soft, porous parts of component *b*, and white representing open parts of this component. Depending on the scale, *a* or *b* may dominate and the matter of REV then becomes important. Thus, focusing on open voids and soft gels at high magnification gives very high F_2 and hence also F_3-values, while focusing on dense aggregates gives very low F_2 and F_3. For smectite-rich clays micrographs with an edge length of at least 30 μm seem to be representative for the larger part of the clay matrix. F_2 and F_3 are related to the average bulk density of the clay as illustrated in Figures 3.32 and 3.33.

From such illustrations, one can derive analytical expressions of the parameters F_2 and F_3 as functions of the bulk density at saturation ρ_{bs} and the gel density ρ_{gs} as given by Eqs (3.1)–(3.4) for MX-80 clay.

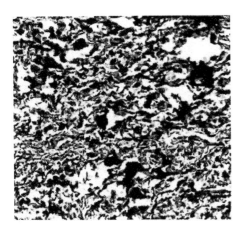

Figure 3.31 Example of digitalized micrograph of Wyoming bentonite (MX-80) with a bulk density at saturation of 1800 kg/m³ (dry density 1270 kg/m³). Black = densest parts of clay matrix *a*. Gray represents component *b* and white represent open parts of this component. Edge length is 3 μm (see Colour Plate II).

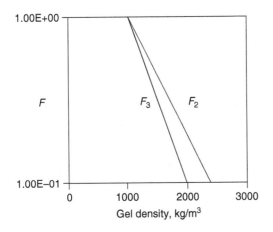

Figure 3.32 F_2 and F_3 versus gel density for Wyoming bentonite (MX-80). Data based on microstructural analyses using 300–400 Å ultramicrotomed sections.

$$F_2 = 10^{(-0.0007875\,\rho_{bs}+0.745)} \tag{3.1}$$

$$F_3 = 10^{(-0.00125\,\rho_{bs}+1.3)} \tag{3.2}$$
$$F_2 = 10^{(-0.0007142\,\rho_{gs}+0.7142)} \tag{3.3}$$

$$F_3 = 10^{(-0.001\,\rho_{gs}+1)} \tag{3.4}$$

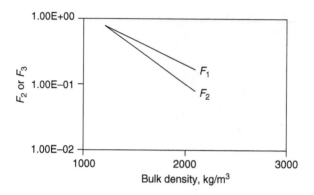

Figure 3.33 F_2 and F_3 versus bulk density for Wyoming bentonite (MX-80).

We will show how these microstructural parameters are related to the hydraulic conductivity and swelling pressure of smectite clay in Chapter 5.

3.5.5   Representativeness

For quantifying microstructural components such as a and b, which represent density variations, the problem is to find out how the section thickness affects the possibility for relevant evaluation. In practice, the issue is to determine how the 'grayness' of the micrograph is related to the density of the specimen. Very dark parts are strongly energy-adsorbing microstructural components such as solid non-smectite minerals, or very dense smectite aggregates. The very bright parts represent open voids filled with impregnation substance. Intermediate degrees of grayness represent clay gels of different density. We will use the schematic representation in Figure 3.34 for this purpose.

Evaluation of grayness

A fundamental law in photography expresses the ratio of grayness G of the photographic plate, that is, the micrograph, and the radiation energy amount E, which is the product of the radiation intensity i and time of exposure t [25]:

$$G = A\,E = \log it \qquad\qquad (3.5)$$

Where G is the grayness, A is the constant, E is the radiation energy, i is the radiation intensity and t is the time of exposure.

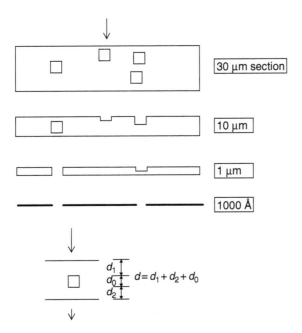

Figure 3.34 The section thickness affects the microstructural interpretation; too thick sections do not reveal fine features. Lowest picture shows a defined element located in the radiation field (light or electron beam) [25].

The grayness G is inversely proportional to the density ρ and thickness d of the film, which, for constant i and t, gives the expression:

$$G = B(\rho d)^{-1} \tag{3.6}$$

where B is the constant, ρ is the density and d is the film thickness.

Disregarding from diffraction, optical distortions and radiation absorption by the void-filling impregnation substance, the grayness of the micrograph intensity on the exit side of the discrete element in Figure 3.34 is:

$$G = B\left[\sum(\rho d)\right]^{-1} \tag{3.7}$$

Comparing a film with 3 μm thickness with one having 30 μm thickness the ratio of the G-values is 10. In order to have direct quantitative comparison of the G-values, the radiation energy must be 10 times higher for the 30 μm film thickness. This may have an impact on the crystal lattice stability, especially the location of alkali ions in EDX analysis.

While identification of the most detailed features of clay crystallites requires a resolution power that only films thinner than 500–1000 Å can provide, certain quantitative microstructural analyses can be made by using significantly thicker films. Thus, if the objective is to interpret different degrees of grayness in terms of density variations, it is possible to do so if calibration has been made by analysing ultrathin sections. For artificially prepared smectite clays using clay powder that hydrates under confined conditions, sufficient information may be obtained by examining 10 μm films. Hence, clays of this sort with a density of 1700–2000 kg/m^3 at water saturation are commonly characterized by the presence of one 3–5 μm wide zone of soft gels per 1000 μm^2 cross-section area. Such a zone located in the centre of the film in the lower sketch in Figure 3.31 would mean that the thickness of the dense matrix $(d_1 + d_2)$ would be about 50% of that of the surrounding homogeneous clay matrix. The grayness of the micrograph where the channel is located would then be 50% of that of the rest of the film. This is sufficient to indicate the presence of such a channel or local void. However, no detailed information on the zone can be obtained and one cannot distinguish between a case with one void and cases with two or several thinner voids located on top of each other. A 30 μm clay film gives even less information on density variations although films with a thickness of about 5 μm provide considerably more information on density variations than 10 μm sections. Micrographs of sections with 1 μm thickness have provided much valuable information although details of the microstructure are obscured.

3.6   References

1. Pusch, R., 1994. *Waste Disposal in Rock. Developments in Geotechnical Engineering*, 76. Elsevier Scientific Publishing Company, Amsterdam.
2. Osipov, V.I., Sokolov, V.N. and Eremeev, V.V., 2004. *Clay Seals of Oil and Gas Deposits*. Balekma Publishing Co., Netherlands.
3. Pusch, R., 1998. Transport of radionuclides in smectite clay. In *Environmental Interaction of Clays*, A. Parker and J.E. Rae (eds), Springer Verlag, Berlin-Hedielberg.
4. Yong, R.N., 2003. 'Influence of microstructural features on water, ion diffusion and transport in clay soils'. *Journal of Applied Clay Science*, Vol. 23: 3–13.
5. Quigley, R.M. and Thompson, C.D. 1966. 'The fabric of anisotropically consolidated sensitive marine clay'. *Canadian Geotechnical Journal*, Vol 3: 61–73.
6. Loiselle, A., Massiera, M. and Sainani, U.R., 1971. 'A study of the cementation bonds of the sensitive clays of the Outardes River region'. *Canadian Geotechnical Journal*, Vol. 8: 479–498.
7. Yong, R.N. and Silvestri, V., 1979. 'Anisotropic behaviour of a sensitive clay'. *Canadian Geotechnical Journal*, Vol. 16: 335–350.
8. Söderblom, R., 1966. 'Chemical aspects of quick-clay formation'. *Journal of Engineering Geology*, Vol. 1: 415–431.
9. Moum, J., Loken, T. and Torrance, J.K., 1971. 'A geochemical investigation of the sensitivity of a normally consolidated clay from Drammen, Norway'. *Géotechnique*, Vol. 21: 329–340.

10. Torrance, J.K., 1975. 'Role of chemistry in the development and behaviour of the sensitive marine clays in Canada and Scandinavia'. *Canadian Geotechnical Journal*, Vol. 12: 326–335.

11. Yong, R.N., Sethi, A.J., Booy, E. and Dascal, O., 1979a. 'Basic characterization and effect of some chemicals on a clay from Outardes 2'. *Journal of Engineering Geology*, Vol. 14: 83–107.

12. Yong, R.N., Sethi, A.J. and LaRochelle, P. 1979b., 'Significance of amorphous material relative to sensitivity in some Champlain clays'. *Canadian Geotechnical Journal*, Vol. 16: 511–520.

13. Kenny, T.C., Moum, J. and Berre, T., 1967. 'An experimental study of bonds in a natural clay'. In the *Proceedings, Geotechnical Conference*, Oslo, Norway, Vol. 1: 65–70.

14. Darban, A.K., 1997. Multi-component transport of heavy metals in clay barriers, PhD thesis, McGill University.

15. Pusch, R. 1970. Clay Microstructure. Swedish National Council for Building Research Document D:8. Swedish National Council for Building Research, Stockholm.

16. Pusch, R., 2002. The Buffer and Backfill Handbook. Technical Report TR-02-20. Swedish Nuclear Fuel and Management Co., SKB, Stockholm.

17. Pusch, R. and Feltham P., 1980. 'A stochastic model of the creep of soils'. *Géotechnique*, Vol. 30: 497–506.

18. Pusch, R., 1973. Influence of organic matter on the geotechnical properties of clays. Swedish National Council for Building Research Document D11:1973. Swedish National Council for Building Research.

19. Watanabe, F., 1970. Personal information on NMR analysis of clays at Tokyo University of Fishery.

20. Pusch, R., 1993. Evolution of Models for Conversion of Smectite to Non-expandable Minerals. Technical Report TR-93-33. Swedish Nuclear Fuel and Management Co., SKB, Stockholm.

21. Nadeau, P.H., Wilton, M.J., McHardy, W.J. and Tait, J.M., 1985. 'The conversion of smectite to illite during diagenesis. Evidence from some illitic clays from bentonites and sandstones'. *Mineral Magazine*, Vol. 49: 393–400.

22. Pusch, R., 1967. 'A technique for investigation of clay microstructure'. *Journal of Microscopie*, Vol. 6: 963–986.

23. Smart, P. and Tovey, N.K., 1981. *Electron Microscopy of Soils and Sediments. Examples.* Clarendon Press, Oxford.

24. Kozaki, T., Tomioka, S., Liu, J., Kozai, N., Suzuki, S., Enoto, T. and Sato, S., 2004. 'Diffusion of radionuclides and microstructure of compacted bentonite observed with micro-CT'. In the *Proceedings from Task Force-related meeting on Buffer & Backfill Modelling*, Lund, Sweden 10–11 March 2004. International Progress IPR-04-23. Swedish Nuclear Fuel and Management Co., SKB, Stockholm.

25. Pusch, R., 1999. 'Experience from preparation and investigation of clay microstructure'. *Engineering Geology*, Vol. 54: 187–194.

26. Pusch, R., Muurinen, A., Lehikoinen, J., Bors, J. and Eriksen, T., 1999. Microstructural and chemical parameters of bentonite as determinants of waste isolation efficiency. Final Report, European Commission Contract No. F14W-CT95–0012.

4   Microstructure of artificial clay-based engineered barriers

4.1   Objective

Smectite clay is widely used for waste isolation, particularly in top and bottom liners of waste landfills. This is primarily because of its excellent sealing ability. The hydraulic conductivity and self-healing capability of the clay are its most valuable properties. Since the cost of high-quality clay is high, mixtures of smectite clay and inexpensive suitably graded soil are often used – as are thin woven mats containing smectite clay of high quality. In recent years smectite clay has become an important engineered barrier material for use in the deep disposal of radioactive waste. This subject will form the focus of this and subsequent chapters. As in the case of natural sedimented deposits of smectites, the macroscopic (bulk) physical properties of artificially prepared smectite clay depend very much on the resultant microstructure obtained in the preparation process. This can be considerably more complex than natural clay sediment deposits.

The focus in this chapter will be on artificially prepared smectite clay for use in the isolation of high level radioactive waste (HLW), and for use as engineered barriers encapsulating hazardous and non-hazardous waste land-fills. All these applications demand a high degree of homogeneity in the clay. This is an implicit requirement for attaining the lowest possible hydraulic conductivity – a primary goal in all kinds of waste isolation schemes.

4.2   Preparation of smectitic barriers

Different techniques are used for preparation of engineered clay barriers, depending on the end-purpose use. For construction of liners, clay at optimum water content is prepared and compacted using compaction tools such as vibrating sheep-foot or pad-foot rollers. For containment of particularly hazardous waste, research has shown that prepared compact blocks of smectitic clay are excellent candidates for use in buffers and barriers for isolating the waste – as shown in Chapter 1. In the first method, compaction of clay material in-place gives a relatively low density. The use of prepared compact blocks as buffer material can produce clay buffers and barriers that are very dense and tight. Microstructural differences between the two end products exist.

4.2.1 Radioactive waste

Highly radioactive waste is contained in metal canisters placed in tunnels and rooms excavated in rock [1]. According to the disposal-management scenarios depicted in most national repository concepts, the canisters will be embedded in dense, artificially prepared smectite clay (the 'buffer') to isolate them from rock movements and groundwater flow. The clay must satisfy strict criteria established in respect to: (a) its rheological performance, (b) its hydraulic properties and characteristics, (c) its gas and heat conductivity, (d) its robust performance characteristics and longevity and (e) composition, that is, it must consist of smectite in a very dense compact state. For connecting tunnels, chambers and other openings that are not the canister depository, backfill with smectitic soil is required. This is necessary to ensure that these tunnels, chambers and non-repository openings will not serve as major hydraulic conductors.

The requirement that the extremely low hydraulic conductivity of the smectite buffer be maintained for periods in the order of hundreds of thousands makes the hydraulic conductivity of the clay its most important property. This requirement is necessary because we need to retard the diffusion transport of errant radionuclides such that they would be harmless when they emerge onto the land surface environment. The required retardation time period ranges between 10 000 and 1 000 000 years – depending on the type of radionuclide involved. It is important to recognize that the host rock for the repository can be crystalline, salt or argillaceous and that the temperatures for the canisters range from 90°C to 150°C. A simple illustrative schematic of the system is shown in Figure 4.1. The 'buffer' consists of the smectitic clay blocks discussed in Chapter 3. These are placed and prepared in a configuration that satisfies the sealing requirement of a buffer. The backfill is obtained by on-site compaction of coarse clay powder or mixtures of clay powder and granular ballast.

4.2.2 Chemical and low-level radioactive waste (landfill)

Chemical waste such as hazardous ash is not contained in long-lasting metal canisters but is compacted in layers and isolated by artificially prepared smectitic clay in top and bottom liners that must meet certain criteria. These conformance and performance criteria are different in different countries [2]. The time for required effective isolation of these types of waste can range from 300 years to several thousands of years. Whilst they are commonly not heat-producing, they have the ability to produce and release gas. The liners are made by on-site compaction of coarse clay powder or mixtures of clay powder and granular ballast. The top liner must be protected from expansion, erosion, drying and frost as well as from damage by gas percolation through a suitably composed overburden. The design of such landfills is very complex. By all accounts, no global, regional or national standards, criteria, conformance and performance requirements,

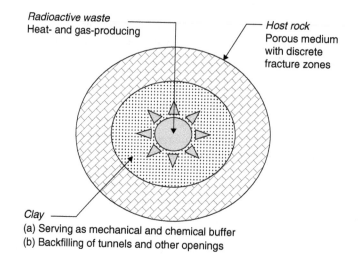

Figure 4.1 System of components in deep disposal of highly radioactive waste.

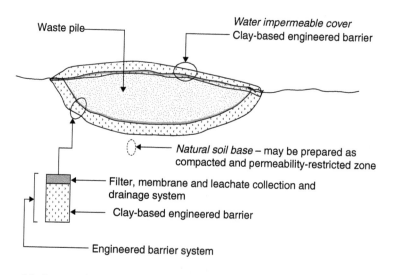

Figure 4.2 System of components in shallow disposal of chemical waste and low-level radioactive waste.

or even basic principles have been agreed upon. In principle the system of components is as shown in Figure 4.2. Another way of effective containment and isolation of hazardous waste from the biosphere is to use abandoned mines in which rooms and drifts are filled with compacted waste using clay lining of walls, roofs and floors [3].

4.3 Microstructural evolution of smectite 'buffer'

4.3.1 General

Artificially prepared smectite clay of required density is produced by compacting smectite clay powder. Water saturation causes expansion of the powder grains. This will lead to a measure of homogeneity of the saturated sample if the sample is confined during the saturation process. At this stage, all we can say is that macroscopic homogeneity is probably obtained. Note that this by no means indicates that the sample is microscopically homogeneous. This is obtained when complete and thorough exfoliation of the stacks of lamellae from the dense grains occurs. A considerable time period is required for this to happen. In this time period, redistribution of internal stress and moisture will be a continual process. Equilibrium in both stress and water content can be achieved between MUs and perhaps in some units themselves.

It is doubtful that complete and thorough exfoliation of all the stacked lamellae in the dense grains and in the MUs will be attained. To a large extent, this is because the artificially prepared smectite will not attain the uniform particle distribution that characterizes natural, dense bentonites. Why? Because the internal resistances to particle redistribution from quasi-frictional forces will effectively prohibit uniform movement and distribution of particles. What this means is that the artificially prepared sample will not be microscopically homogeneous. This fact, which is obvious from microstructural modelling, explains the difference in physical properties between different deposits, samples and types of bentonites.

4.3.2 Raw material

The raw material, usually obtained from a bentonite bed commonly used for preparation of commercial smectite-rich clay is crushed and dried in rotating kilns. The material is subsequently ground to yield clay powder with the desired size distribution as shown in Figure 4.3. Since most raw

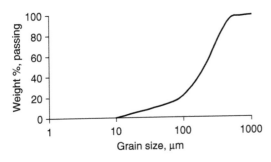

Figure 4.3 Typical grain size distribution of American Colloid's MX-80.

smectite materials are in the Ca form, and since a Na-smectite is the preferred material, conversion to the Na state requires the addition of a small percentage of sodium carbonate to the clay powder. The cation-exchange process that will occur over a period of weeks and/or months is associated with the formation of $CaCO_3$ that can have a cementing function.

4.3.3 Constitution of powder grains

Commonly exploited bentonites have a dry density of 1900–2200 kg/m³ with 1 or 2 hydrates in the interlamellar space. Note that for practical reasons, one can take the weight of the clay to include the hygroscopic water, which is about 10% of the clay mass if it is stored at a relative humidity (RH) of 40–60%.

To manufacture blocks consisting of particles of smectite-rich clay (e.g. commercial bentonites) one can use ordinary crushed and air-dried material. This will generally have a water content of from 8% to 12% when stored at 50–70% RH. The grains are composed of numerous montmoril-lonite lamellae, which form stacks and aggregates of stacks as basic units. The stacks are typically 3–10 nm thick with diameters of about 0.05–0.25 μm. Individual aggregates can be several hundred μm. When processing of the material is complete, it is found that these stick together to form particles of several mm to cm in size (Figure 4.4). Wetting of the clay grains in the compacted block is necessary to develop proper thermal conductivity properties. There is a definite balance between the degree of saturation reached with water uptake and the workability of the wetted blocks. At the one end, one requires compact blocks at near saturation for optimal thermal conductivity properties. However, the confining pressures needed to allow this

Figure 4.4 Scanning electron micrograph (SEM) of a lump of particles in dried and ground natural bentonite (Wyoming, MX-80). The edge length of the little block is about 5 mm.

to happen are more than the pressure used to obtain the compacted blocks. Expansion of the compacted blocks upon water uptake is not a desirable feature at this stage. Experience has shown that one should not exceed water contents of about 15–20%.

4.3.4   Compression and hydration of powder grains

Air-dry clay powder is poured in a form and compacted uniaxially or triaxially under high pressure to form highly compacted blocks. A common procedure for manufacturing clay blocks by compacting air-dry clay powder in a form, uniaxially under a pressure of 50–150 MPa. The dry density of the grains is about 2000 kg/m^3 for 10% water content by weight, and the dry density of the powder mass poured in the form is about 1200 kg/m^3. Uniaxial compression under this pressure causes a reduction in void space, and deformation of the grains leading to a dry density of about 1850 kg/m^3 and a density after complete water saturation of about 2050 kg/m^3.

Figure 4.5 shows a 'buffer' block prepared by compaction of commercial bentonite powder for embedment of a canister with highly radioactive waste, and Figures 4.6 and 4.7 shows blocks for use in half-scale 'mock-up' experiments.

The compaction and maturation processes can be simulated by applying suitable numerical tools such as the boundary element code BEASY [4], assuming the powder grains with 10% water content to behave elastically according to experimentally determined stress/strain relationships.

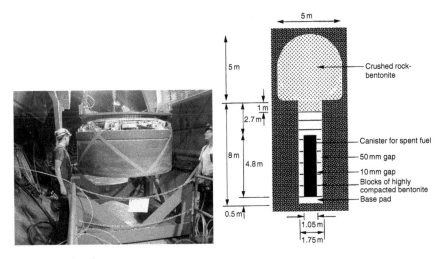

Figure 4.5 Block with 1.85 m diameter and 2000 kg weight prepared by compacting Wyoming bentonite (MX-80) powder under 100 MPa uniaxial pressure. The block is being placed into a deposition hole in line with the buffer objective shown in the concept drawing in the figure on the right [1].

Figure 4.6 Clay block prepared by compacting clay powder under 100 MPa pressure. Block with 0.3 m diameter of Greek Ca bentonite powder converted to sodium form. Uniaxially compressed blocks tend to fracture perpendicular to the pressure direction.

Figure 4.7 Sector-shaped block of 85% smectite-rich clay (Fe-montmorillonite/vermiculate) mixed with 10% quartz and 5% graphite to enhance heat conductivity. Compaction pressure of the powder mix was 100 MPa pressure. Despite the small graphite content the block is almost black.

Considering a unit cell of the powder consisting of one big grain with 0.35 mm diameter and 8 small grains with 0.10 mm diameter (Figure 4.8), which roughly matches the size distribution of common clay powders, the compressive uniaxial strain caused by 100 MPa pressure is found to be about the same as in full-scale compaction, that is, about 30%.

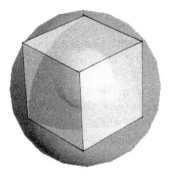

Figure 4.8 Unit cell with 1/8 of big grain (0.35 mm) contacting one small grain (0.10 mm). The edge length of the cubical unit cell is 0.175 mm.

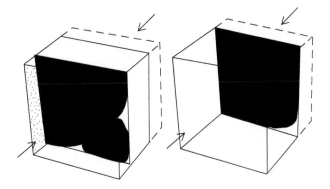

Figure 4.9 Model of the unit cell compressed in the compaction phase and then hydrated resulting in permeable channels. Two sections through the hydrated cell are shown with the black colour representing the dense clay matrix [5].

Theoretical and model calculations show that the stresses are very high in the centre of the grains and at their contacts. The Mises stress exceeds 200 MPa in the interior of the grains from where interlamellar water is squeezed out to the less loaded shallow parts of the grains. Subsequent uptake of water from the boundaries of the unit cell causes expansion of the grains. This process will be different in different parts of the grains: the dense central parts being under high pressure will take up little or no water, while the rest will expand. Taking the expansion to be exclusively caused by interlamellar uptake of water, one can calculate the resulting 'external' porosity and pore dimensions. This has been made by the use of BEASY [4]. Figure 4.9 shows the outcome of the calculations in simplified form.

The unit cell compressed uniaxially in the compaction phase is shortened and subsequent hydration results in expansion of the grains. The outcome of this process is such that the remaining voids form tortuous channels with varying apertures. Although there are no straight open channels between the expanded grains through the clay, tortuous channels are formed by interconnected voids with diameters as low as a few μm. All cross-sections of the unit cell of any clay powder with a size distribution similar to that in Figure 4.3, and compressed under 100 MPa pressure, will have an average dry density of about 1850 kg/m^3. Furthermore, these cross-sections will contain about three voids with an average diameter of 10 μm each. They are continuous in the unit cell but may not be so throughout a series of stacked unit cells. Naturally, there are also many isolated voids of smaller sizes.

The very high shear stresses that are associated with the compaction exceed the shear strength of the grains. This results in fragmentation of some grains. The fragments will move into the voids where they will rearrange and form gels of lower density than the central parts of the grains. This will result in considerable variations in density on the micro scale, and testifies to the significance and importance of the role of MUs on the properties and characteristics of the smectites in the compacted blocks.

4.3.5   Gel formation processes

The hydration of the pressurized powder grains generates shear stresses that fragment numerous stacks of lamellae resulting in exfoliation from the grains. As stated in the previous section, the released clay fragments are free to reorganize and form gels in the voids and channels. Their density will be controlled by the porewater electrolytes (Figures 4.10 and 4.11).

Figure 4.10 Schematic picture of coagulation of clay particles. Left: structurally homogeneous clay network. Right: coagulation by increased electrolyte concentration in the porewater.

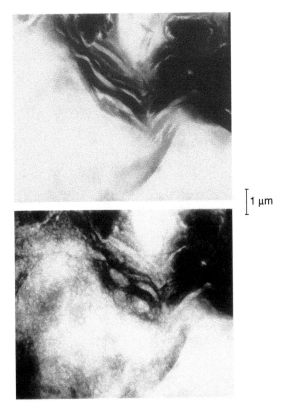

1 μm

Figure 4.11 Expansion of Na-smectite clay by exposure to water vapour in a CNRS high-voltage electron microscope [6].

This process is of great practical importance in situations such as plugging of deep boreholes with highly compacted clay. It can be achieved by applying the technique shown in Figure 4.12. Perforated corrosion-resistant tubes with compacted bentonite blocks are lowered into steep boreholes or pushed into the boreholes in any desired direction. Figure 4.13 shows that the front of the clay that expands through the perforations, and eventually embeds the tubes with a nearly homogeneous dense clay, is a very soft clay gel. This gel is later consolidated under the swelling pressure exerted by the clay that is still in the tube.

4.3.6 Ultimate microstructure of highly compacted smectite 'buffer' clay, matured under confined conditions

An important conclusion from the large number of microstructural investigations that have been made on artificially prepared clay seals is that

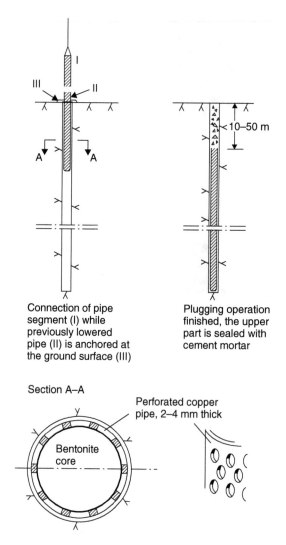

Figure 4.12 Procedure for borehole plugging with compacted clay blocks [1].

the homogeneity of a confined system of expanded grains is improved by the gels formed in the voids. However, the basic microstructural pattern of powder grains persists. Hence, natural clays are far more homogeneous than artificially prepared ones. This is of special importance for gas permeation.

Compaction of powder grain assembly

As described in the previous sections, compaction of bentonite powder grains deforms them. Furthermore, the variation in size and position of the

Figure 4.13 Growth of soft clay through the perforation of an 80 mm copper tube with dense bentonite in an oedometer – to simulate maturation of a borehole plug. After 8 hours, the central part of the bentonite core is still unaffected by water. After a few days, the soft gel is densified by clay moving from the core through the perforations (see Colour Plate IX).

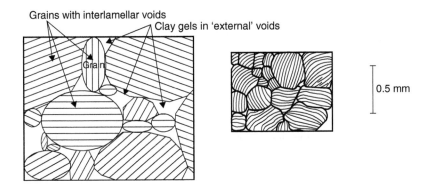

Figure 4.14 Arrangement of granules in compacted MX-80. Left: low dry density with voids between granules measuring up to a few hundred μm wide. Right: dense condition at 100 MPa compaction with voids up to 50 μm wide. The granules contain numerous small isolated voids [1, 5].

grains mean that the grains are brought into dense layering with voids of varying sizes. For compacted bentonite with low densities, the voids form continuous channels with constrictions. In the case of the dense compacted samples, the voids tend to be isolated – as shown in Figure 4.14. Studies with respect to microstructural modelling of compacted clay granules use the range of void sizes for densely compacted MX-80 clay in Table 4.1.

Table 4.1 Void size distribution in MX-80 with 10% water content compacted to 1850 kg/m^3 dry density

Void size, μm	Percentage of voids
<50	100
<7.5	75
<3.5	50
<2	25

Ultimate state after hydration and maturation

We recall from previous discussions that smectite particles consist of stacks of lamellae that are separated by hydrated cations or positively charged molecules. Their size and charge and interaction with water molecules determine the interlamellar spacing. The numbers of lamellae are different for Na (3–5) and Ca smectite clays (about 10). In natural sediments the stacks of lamellae, which are oriented more or less perpendicularly to the direction of the overburden pressure, form an interwoven network of apparent continuity. In artificially compacted clay however, orientation of the stacked lamellae is usually random. This is because the individual particle aggregates with their inherited internal parallelism are arbitrarily oriented in the compaction process (Figure 4.14). Hydration of the aggregates results in swelling and physical shearing of particles, the outcome of which is significant internal movements of the particles. At elevated temperature, the combination of these complex internal movements with the formation of gels by exfoliated stacks of flakes will ultimately produce a mature clay that is macroscopically isotropic. This means to say that in the ultimate mature state, the clay exhibits bulk isotropic properties.

The physical properties of bulk artificially prepared smectite clays are determined by the same basic sets of forces and particle interaction mechanisms that control the behaviour of natural sediments. An example of this is shown in Figure 4.15 with respect to the properties of hydraulic conductivity and swelling. The only major difference is the gas permeability, which is believed to be lower for natural smectite sediments, hence explaining the confinement of pressurized gas in deep oil- and gas-bearing sediments. One should notice the impact of density on the swelling pressure, particularly for Ca-montmorillonite. Thus, for bulk densities at complete fluid saturation lower than 1600 kg/m^3 there is practically no swelling pressure.

Quantitative microstructural description of clay 'buffer'

The microstructural *F*-parameters, discussed in Section 3.5.4, can be used to describe the quantitative relationship of the microstructure to the transport

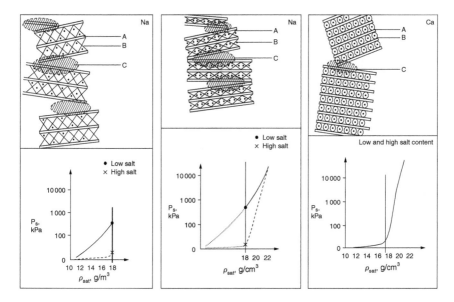

Figure 4.15 Schematic pictures of stack assemblages and influence of density at water saturation, expressed in g/cm³ (1 g/cm³ equals 1000 kg/m³) and salinity for Na and Ca montmorillonite clay. (A) lamella, (B) interlamellar space, (C) contact region with interacting electrical double-layers [1].

Table 4.2 Microstructural data for MX-80 in Na form (F_2-value)

Bulk density, kg/m³ at saturation	Dry bulk density, kg/m³	Gel density, kg/m³	F_2
2130	1790	2000	0.17
1850	1250	1650	0.24
1570	905	1150	0.40

and rheological properties of natural and artificially prepared clays – as will be shown in Chapter 5. In this section, we will confine ourselves to the provision of parameter data and to a description of a microstructural model developed for heterogeneous clays such as the ones prepared by compaction of clay powder.

The results of comprehensive studies of electron transmission micrographs have provided data relating to the fraction F_2 of a section that is permeable. Examples of these are given in Table 4.2 for Wyoming bentonite (MX-80) as a function of the bulk density. The table also contains the evaluated average density of the clay gel contained in the channels between dense, impermeable particles. Data are given for three typical bulk densities representing 'buffer' clay. For the highest density, the F_2 value means that only 17% of a cross-section is permeable and that water flows through a clay mass with an average density at water saturation of 2000 kg/m³, that is, almost the same as the average bulk density. Since the permeable part of the clay matrix also represents the part of the clay in which pore diffusion takes place, F_2 is also a measure of the diffusive transport capacity.

For artificially prepared smectite-based clays, expandability is of great practical importance since it: (a) determines the tightness of contacts with rock and waste canisters and (b) establishes the nature of the support that dense clay can provide to the rock where clay seals are placed. The swelling pressure of the clays is the physical parameter of interest. It is a direct function of the density of the clay – as indicated in Figure 4.15.

While the hydraulic and gas conductivities and also the diffusive transport capacity are related to the F_2 parameter, the swelling pressure engages the particle network in all directions. This calls for the use of the F_3 parameter. Evaluation of micrographs has given data for this parameter – as listed in Table 4.3 for Wyoming bentonite (MX-80).

Influence of ion exchange from Na to Ca

With respect to the influence of ion exchange, it is noted that only a small change in the internal force fields results when exchange from Na to Ca for the highest density occurs. In contrast, a change from Na to Ca of low-density clays causes contraction of stacks and coagulation of the gels, leading to a strong increase in hydraulic conductivity and drop in swelling pressure. We will see in the subsequent chapter that this process actually occurs in laboratory tests.

Influence of interlamellar sorption of large organic molecules

Mixing of clay powder with organic material for certain purposes such as enhancement of the anion-exchange capacity significantly changes the microstructure. Correspondingly, there will be changes in the F-parameter data. We will illustrate this by a case study with (HDPy⁺) involving Wyoming bentonite (MX-80) [5]. The study involved dispersing Wyoming bentonite (MX-80) in water, and where a chloride salt of the quaternary alkylammonium ion of hexadecylpyridinium (HDPy⁺) was added to obtain complete cation exchange from the original Na/Ca/Mg-state. HDPy⁺ was

Table 4.3 Microstructural data for MX-80 in Na form (F_3-value)

Bulk density, kg/m³ at saturation	Dry bulk density, kg/m³	Gel density, kg/m³	F_3
2130	1790	2000	0.07
1850	1250	1650	0.20
1570	905	1150	0.25

adsorbed to about 150% of the cation-exchange capacity (CEC). However, it is estimated that the process occurred in cationic form up to 100% CEC, and that additional uptake was in molecular form. After drying, the powder was compacted under pressures up to 200 MPa.

The sample was subsequently saturated with distilled water to allow for determination of its physical properties. This was followed by extraction of specimens for acrylate-treatment and preparation of 400 Å sections for transmission electron microscopy (TEM). For comparison, 10 μm sections were prepared by grinding and polishing according to the common accepted procedure in petrology. The evaluated *F*-parameters, which are given in Table 4.4, reinforces the earlier statements made in the previous chapter that thicker sections conceal many small voids. Figure 4.16 illustrates the microstructural heterogeneity of the organic form of MX-80. This feature is in strong contrast to ordinary organic-free clay of the same density. We will show in Chapter 5 that it explains why higher hydraulic conductivities are generally obtained with clays treated in this fashion.

Table 4.4 Microstructural parameters of HDPy$^+$-treated MX-80 with a density of 1760 kg/m^3 at saturation with distilled water

Micrograph type	F_2	F_3
TEM	0.51	0.37
Light microscopy	0.29	0.16

Figure 4.16 Digitalized TEM micrograph of HDPy$^+$-treated MX-80 clay with 1760 kg/m^3 density in water saturated form. The width of the micrograph is 1 mm. Wide voids forming channels with up to 100 μm width separate the dense and homogeneous blocks of aggregates [5] (see Colour Plate III).

Treatment of smectite clay with organics has been found to be useful: (a) for all types of pelletization, (b) for improvement of the performance of the treated clay as a drilling mud and particularly (c) for application in applied microbiology, cosmetics and pharmacology. In view of the impact of such treatment on the material structure and properties, it is important that a proper understanding and appreciation of the microstructure of the treated material be obtained. Although considerable scientific studies are being conducted globally in many scientific disciplines, the problem for manufacturers is to obtain a sufficiently homogeneous mixture of clay and additives under field conditions. Although homogeneity can be readily obtained in the laboratory environment, such has not been the case for the field environment. As will be seen in the following section, the problem of preparation of mixtures of different types and sizes of soil is a significant issue.

4.4   Microstructural evolution of smectitic fills

4.4.1   General

Although smectite-rich clays offer better isolation capability for containment of waste, and for more effective sealing of tunnels, shafts and waste landfills, for economic reasons, relatively smectite-poor natural clays and even mixtures of smectite and ballast (aggregate) material are often used. Another reason why smectite-poor natural clays are used is because of the high swelling potential of compacted smectite-rich clays. To obtain low hydraulic conductivities, significantly large confining pressures are needed to counter the swelling pressures. These kinds of pressures are not easily found in the types of engineered barriers now being used, that is, top and bottom liner systems.

Since the microstructure of smectite-poor engineered barriers is influential in the control of their physical properties, we need to consider the use of such materials in the design of barrier containment systems. For simplicity and clarification in the use of terms, we have chosen to call *fills* those clay materials that contain less than 50% *smectitic clay*. The term *ballast* is used to indicate non-swelling soil particles, that is, essentially representing crystalline rock.

4.4.2   Mixtures of smectite-rich clay and ballast

Basic cases

Mixtures of clay and ballast with different compositions can be specified and obtained to meet desired performance requirements, that is, low hydraulic conductivity and low compressibility. A basic principle that one could follow to obtain material that would meet the requirements is to create a mixture such that the required tight spaces between non-clay ballast are occupied by clay minerals [1, 5, 7]. When such a mixture is obtained, the hydraulic conductivity of the mixture will be low. This is especially true if the clay contains smectite minerals. The compressibility of the mixture is low because the skeletal structure is composed of ballast grains. These will carry almost the entire externally applied load. This same skeletal structure, with tight pore spaces between the ballast grains, does not allow for much swelling of the pore-filling clays to occur – because of the structural strength of the skeleton. This principle is demonstrated in Figure 4.17. The schematic diagram shows the difference between: (a) the basic case with only ballast material (top sketch) (b) with the ballast material mixed with smectite clay particles (middle sketch) and (c) swelling of the smectite clay particles in the voids to form a low-permeable clay gel (bottom sketch).

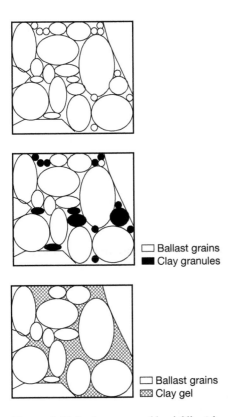

□ Ballast grains
■ Clay granules

□ Ballast grains
▨ Clay gel

Figure 4.17 Performance of backfill without and with clay component. Upper: basic case with only ballast material. Centre: case with smectite clay particles in the voids between larger ballast grains. Lower: previous case with the clay hydrated and expanded to form a clay gel in the voids [7].

The proportion of clay content is very important. As in other geotechnical contexts, a clay content exceeding about 15% means that the clay matrix becomes continuous. This will result in a situation where the ballast grains are effectively 'floating' in the clay matrix. When this occurs, the properties of the clay will dominate. This is both good and bad. It is good for hydraulic conductivity – provided that the mixture can be compacted to a very dense state. It is bad, however, because of the swelling potential and the deformability of the mixture. For a sensible mixture that would allow for a proper development of a skeletal structure and optimum performance with respect to bulk hydraulic conductivity and compressibility, a Fuller-type ballast grain size distribution with 5–10% swelling clay by weight can be used (Figure 4.17).

The swelling clay that is commonly used is a commercial, finely ground smectite-rich Na bentonite with montmorillonite as major smectite species. In practice, the clay content should be somewhat higher than the theoretical 5–10% range. Generally, it has been found that because most bentonites do not possess more than about 70–85% smectite (e.g. MX-80), one needs to use a mixture where the proportion of bentonite is about 10–15%.

Several problems are encountered with low clay contents. The most important ones are:

- Even effective compaction of the mixture, the density of the clay component cannot be increased sufficiently to meet the very low hydraulic conductivity requirements. This is because the skeletal structure formed by the ballast grains bears most of the compactive effort exerted on the mixture.
- The degree (percentage) of void-filling by the clay can be low.
- The sensitivity to mechanical strain, for example, uneven settlement of a landfill top liner, is strong.

Whereas cost is always a key factor, leading to use of the cheapest possible materials and application techniques, this should not obviate the need for comprehensive material testing for quality assurance.

Importance of fines in the ballast

The granular composition of both ballast and clay particles are important factors in the distribution and density of the respective components. Recent experience has shown that the fine fractions of the ballast material contribute significantly to the development of low hydraulic conductivity in the material. This is not unexpected since the fines will populate the voids and contribute to *void plugging*. This subject will be discussed in a later chapter. The commonly used geotechnical size distribution parameter $C_u = D_{60}/D_{10}$, is not a determinant of the hydraulic conductivity. However, the D_5 is a key parameter – testifying to the fact that one only needs a small percentage of the finest fraction of the ballast material to markedly affect the hydraulic conductivity.

Microstructural modelling

Attempts have been made to model the homogenization process of clay/ballast mixtures using a general material model [8]. The study, which was performed in 2D for mixtures with about 10% clay, was based on the assumption that the clay particles and ballast grains were rod-shaped and regularly arranged as shown in Figure 4.18. The ballast grains were assumed to have a diameter of 2.5 mm while the bentonite grains were taken to be 1 mm in diameter, both roughly corresponding to the median size of experimentally investigated mixtures.

Given the two grain sizes the microstructural model geometry represents the densest possible layering. The dry density of the granules was set at 1980 kg/m^3 whilst that of the ballast grains was taken to be 2650 kg/m^3. The dry density of the mixture was 1560 kg/m^3, corresponding to 2000 kg/m^3 at water saturation. Figure 4.18 also shows successive stages in the homogenization process. The process ultimately yields a very significant variation in clay dry density, that is, from 620 to 1450 kg/m^3, corresponding to a density at water saturation of 1400–1930 kg/m^3 [8].

The study indicates that the narrowest portion of the voids between the ballast grains will not be filled by the clay fraction. The resulting open space, which forms isolated unfilled ring-shaped zones around the contact points of ballast grains has some effect on the overall hydraulic conductivity. However, the heterogeneity of the clay is more important. Clay density will be low in the most expanded part of the clay grains, resulting in the development of high hydraulic conductivity. The risk of piping and erosion of the softest parts will be considerable.

The quantitative data from the study are not entirely realistic, primarily because the model is not in 3D. Nevertheless, the results obtained are of importance from the point of impact of physical chemistry. They suggest very strongly that the softer parts of the clay occupying the space between the ballast grains will undergo coagulation and become very permeable if the groundwater is saline. This is consistent with observations of the difference between the hydraulic conductivity values of backfills constructed with coarse-grained clay components and backfills constructed with finer-grained clays. It is evident that the prudent course of action is to use very fine-grained clay when the amount of clay added to the ballast is small.

The impact of the grain size distribution of both ballast and clay particles indicates that both types of materials should conform to the Fuller or Weymouth parabolic curves to minimize the size and volume of unfilled voids. Figure 4.19 shows the curves in the original form published by the United States Bureau of Soils Classification.

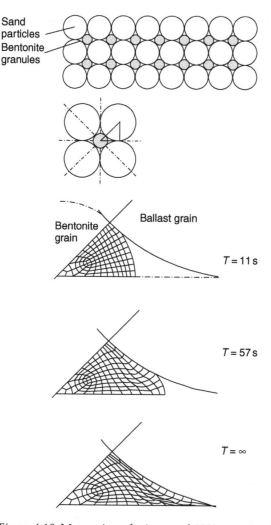

Figure 4.18 Maturation of mixture of 10% smectite clay (MX-80) and ballast particles according to a finite element (FEM) analysis in 2D [6]. The lower pictures show the expansion of the clay between the ballast grains. The narrow gel in the ultimate stage has a density at saturation of 1420 kg/m³, while it is 1930 kg/m³ for the densest part (*T* is time in seconds) [8].

Semi-empirical modelling

Figure 4.20 shows the relationship between bentonite content and mixture dry density for a backfill compacted to 100% Standard Proctor density. The results portrayed in the figure show that the mixture can have a dry density as high as 2000 kg/m³ for a bentonite content of less than about 20%.

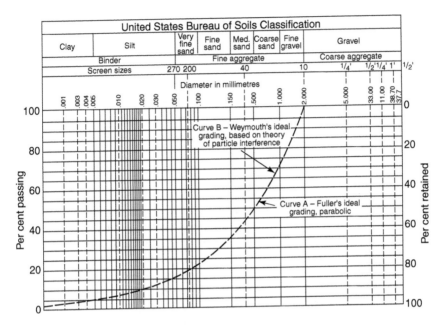

Figure 4.19 Fuller's ideal grading (parabolic curve) almost coinciding with Weymouth's ideal grading [7].

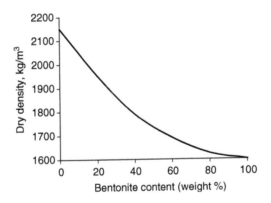

Figure 4.20 Relationship between bentonite (MX-80) content and dry density of the mixture at 100% Proctor compaction. BMT ballast.

Using fundamental soil physical relationships, one can express the porosity n_s of the ballast as in Eq. (4.1) [7]:

$$n_s = 1 - \rho_d/\rho_s(1 - b) \tag{4.1}$$

where ρ_d is the dry density of the mixture, ρ_s is the density of the minerals and b is the bentonite content (weight fraction).

A certain fraction of the ballast voids are filled with bentonite. If the density of the clay filling is ρ_{db} the degree of clay filling S_b of the voids between ballast grains will be as specified by Eq. (4.2):

$$S_b = [\rho_s/\rho_{db}]/[(\rho_s/\rho_d)1/b - (1/b + 1)] \qquad (4.2)$$

Assuming uniform densities and equal compactive efforts for the mixtures and clay fractions, and taking the maximum density at water saturation of the clay fraction to be 1600 kg/m³, the relationships in Table 4.5 for different clay contents can be derived.

Influence of the type of ballast material

As in the case of the importance of type of aggregates and gradation and the quality of the resultant concrete obtained, the type of ballast (aggregate) plays a major role in the performance of clay/ballast mixtures. Ballast grains with low strength can be severely degraded by heavy compaction during construction of backfills and top liners. The fragments, together with their fragmented parents do not have the ability to seek optimum rearrangement and distribution – the end result of which is, more often than not, a less-than-stable compacted mass. A better and more physically stable ballast material will be glacial sediments with rounded shape of the grains since they have the ability to move more easily into stable positions upon compaction, resulting thereby in the production of higher densities. Although crushed rock may be the cheapest ballast material obtainable in large quantities, two simple factors need to be considered: (a) the ability of individual aggregates to move easily into stable positions under compaction load and (b) the affinity of debris, such as very small fragments of rock-forming

Table 4.5 Semi-theoretical relationships between clay contents, compaction degrees and physical soil data [7]

Clay content, %	Degree of Proctor density, %	ρ_d, kg/m³	ρ_{sat}, kg/m³	S_b	n_s	$\rho_b{}^a$, kg/m³
10	100	2050	2290	0.40	0.317	1600
10	87.5	1790	2130	0.32	0.403	1400
10	75	1540	1970	0.25	0.487	1200
20	100	1950	2230	0.57	0.42	1600
20	87.5	1710	2070	0.50	0.49	1400
20	75	1460	1920	0.43	0.57	1200
50	100	1750	2100	0.81	0.67	1600
50	87.5	1530	1960	0.76	0.72	1400
50	75	1310	1830	0.72	0.76	1200

Note
a Density of clay content at complete water saturation.

minerals, for the angular rock fragments. One needs to avoid creating less-than-clean surfaces of the individual aggregates. The debris presence on the surfaces will prevent the clay from adhering to the grains. They have the potential for development of flow paths in the backfill (Figure 4.21).

4.4.3 *Smectite-poor natural clays for filling purposes*

Natural smectitic clays that are homogeneous and consisting of very fine-grained material, have several advantages over that of mixtures of clay and ballast. The most obvious beneficial property is that practically all grains contain some smectite. Furthermore, the compacted material will have a more homogeneous microstructure in comparison to mixtures. These advantages make the natural smectitic clays good candidates for use in tunnels or as bottom and top liners. The lower picture in Figure 4.17 is a good example of the difference in microstructural constitution of mixtures and natural clayey materials, assuming the ballast grains to contain some smectite.

To all intents and purposes, almost every grain in the natural clay material produces clay gels, whereas numerous voids are devoid of clay in the mixtures. The necessary conditions wherein natural smectitic soils can be used for construction of engineered barriers are: (a) that sufficiently large amounts of uniform material can be found and exploited and (b) that they must not contain minerals or other species such as organics that will allow for gas production or cementation. As an example, certain Triassic clays

Figure 4.21 Micrograph of thin section of acrylate-embedded mixture of 10% MX-80 bentonite and 90% crushed TBM muck. The picture shows coatings (black) of the rock fragments by a thin layer of very small quartz particles, which strongly increase the permeability of the mixture. Magnification 100× [5] (see Colour Plate I).

that have been shown to perform well for waste isolation in laboratory studies, that is, short-term performance studies, will have several problems in the long-term because of their adverse chemical reactions with air or water.

The same basic laws with respect to hydraulic conductivity, swelling pressure and compressibility apply to both natural and mixed fills. The major differences are that a significant amount of clay is required for the mixture if it is to obtain performance equal to that of the natural material. The added cost of mixing in the production of the mixture is a factor of some concern for large projects. The importance of the microstructure of clays in construction of liners will be discussed in a later chapter.

4.5   References

1. Pusch, R., 1994. *Waste Disposal in Rock. Developments in Geotechnical Engineering*, 75. Elsevier Science Publishing Company, Amsterdam-Oxford-New York.
2. Pusch, R. and Kihl, A., 2004. Percolation of clay liners of ash landfills in short and long time perspectives. *Waste Management and Research*, Vol. 22, No. 2 (2204): 71–77.
3. Pusch, R., Schomburg, J., Kaliampakos, D., Adey, R. and Popov, V., 2004. Low Risk Deposition Technology. Final Report, European Commission Contract No. EVG1-CT-2000-00020, Brussels.
4. BEASY, 1995. User Guide, Computational Mechanics BEASY Ltd, Southampton, UK.
5. Pusch, R., Muurinen, A., Lehikoinen, J., Bors, J. and Eriksen, T., 1999. Microstructural and Chemical Parameters of Bentonite as Determinants of Waste Isolation Efficiency. Final Report, European Commission Contract No. F14W-CT95.0012.
6. Pusch, R., 1987. 'Identification of Na-smectite hydration by use of "humid cell" High voltage microscopy'. *Applied Clay Science*, Vol. 2: 343–352.
7. Pusch, R., 1992. The Buffer and Backfill Handbook. Swedish Nuclear Fuel and Waste Management AB (SKB), Technical Report TR-02-12. Stockholm.
8. Börgesson, L., 1993. Swelling and Homogenisation of Bentonite Granules in Buffer and Backfill. Swedish Nuclear Fuel and Waste Management AB (SKB). AR 95-22. Stockholm.

5 Clay properties and microstructural constitution

5.1 Objective

The bulk behaviour of a clay is the overt manifestation of the sum of the interactions of all of the microstructural units (MUs) that comprise the macroscopic skeletal network of the clay. We define the *microstructural constitution* of a soil to mean a soil that is composed of MUs that combine to form the macroscopic skeletal network of the soil. The mechanical, physical and physico-chemical properties of a clay soil are direct products of: (a) the homogeneity or lack thereof (heterogeneity) of the microstructural constitution and (b) the reactions between MUs themselves and between the MUs and the porewater. The properties and interactions of interest such as hydraulic conductivity, gas penetrability and ion diffusion, transport, shear strength, creep and erodability will be considered in this chapter. Our interest centres around the establishment of meaningful relationships between the microstructural constitution of clays and the aforementioned properties – with particular focus on smectitic clays.

5.2 Hydraulic conductivity

5.2.1 General

Variations in the nature, number and distribution of the MUs in a clay soil are reflected as density variations. From a purely physical point of view, these density variations are indicative of the changes in the geometrical arrangements of the MUs, that is, the fabric of the soil. We recall from the previous chapters that we have defined the *fabric* of a soil to mean the geometrical arrangement of the soil fractions, including particles and MUs. We can also recall that we have defined the *structure* of a soil to include the forces and interactions between those particles and MUs that constitute the soil. In simple terms, we can consider the *soil structure* to be soil fabric with forces and interparticle actions.

Clay soil fabric has a very significant influence on the hydraulic conductivity of the soil. To a very large extent, this is because of the voids, their interconnectivity, their dimensions and the tortuosity of the flow paths. The limiting dimensions or sizes of the flow paths are obviously critical factors since they control the volume of flow that can be conducted under various conditions. They are extremely small and with limited continuity in very smectite-rich clay such as high density MX-80. For low densities, and particularly when the dominant exchangeable cations are Ca, hydraulic conductivity is measurably increased due to the development of channels. The type of clay mineral plays a dominant role in the mobility of water in and through the soil. Figure 5.1 is a schematic illustration which shows the proportions of water associated with the microstructures and the voids separating MUs. The hydration water (hydrates) within the interlamellar spaces, combined with the water in the micropores of the microstructure, constitute water that is relatively immobile. Water in the macropores and other void spaces have properties that are similar to that of bulk water, and that are relatively mobile. Figure 5.1 shows that much less porewater is immobile in illitic and kaolinite-rich clays – as opposed to that in smectite clay with the same bulk density. Further implications and impacts arising from this proportioning of the more-or-less mobile water can be found in Section 8.3.6 where the microstructure controls on transmissibility and swelling properties of the clay soils are discussed.

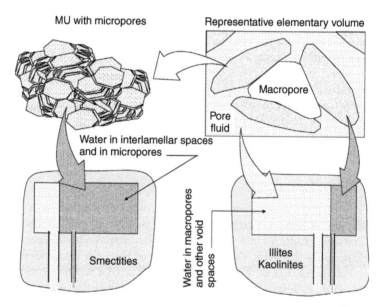

Figure 5.1 Schematic illustration of the proportions of water held in interlamellar spaces, micropores and macropores. The two diagrams below show the respective proportions of water from the sources identified in the top right-hand schematic. Water held in the micropores and interlamellar spaces are significantly less mobile than water held in the macropores and other void spaces.

The geometrical features on the microstructural scale of all sorts of clay can be evaluated from ultrathin sections of suitably prepared clay samples as outlined in Chapter 3. In principle, the denser and softer parts of the clay matrix can be distinguished quantitatively by analysing digital micrographs of the ultrathin sections. By assigning to them characteristic average hydraulic conductivities, the net average conductivity of a clay element can be calculated. Such analyses have provided good agreement between microstructural parameters and the conductivity of bulk clay. This will be discussed further in the latter part of this chapter.

5.2.2 Determination of the hydraulic conductivity of clay samples

The common procedure for determination of the hydraulic conductivity of soils is to use permeameters – either under constant head or falling head conditions. Although there is still considerable debate on the sizes of permeameters and nature of sample confinement, the procedures for test implementation are well documented in the many standards of most countries (e.g. ASTM, EC standards). However, for situations where the soil has a high swelling potential or high swelling pressure upon wetting, the standard procedures need to be modified or even changed drastically. Because of the high swelling pressures that can develop, rigid confinement is required if the cross-sectional area presented to the permeating fluid is to be preserved.

A popular method for determination of the hydraulic conductivity of swelling soils, which has also been applied to non-swelling soils, uses oedometers instead of permeameters. The robust confinement afforded by the oedometer against the swelling pressure developed during percolation (i.e. fluid permeation of the test sample), is the strong feature of this testing technique. Figure 5.2 shows a typical oedometer system. Although the dimensions shown in the diagram are representative of some of the typical systems used, there is no hard and fast rule as to dimensions except that: (a) the cross-sectional area of the test sample must be sufficiently large so as to provide a representative portion of the sample and to avoid testing anomalies and (b) the height of sample should be sufficiently high so as to permit proper hydraulic transport of the permeating fluid.

Darcy's hydraulic conductivity coefficient K

The apparatus shown in Figure 5.2 is equipped with a piston-pressure cell arrangement to allow for the recording of swelling pressures in the direction of fluid permeation. As with the results obtained in the classical permeameter tests, Darcy's model is used for data analysis. In common with other 'flux law' models (e.g. Fick), the compliance K, the coefficient of hydraulic conductivity relates the flux rate given as the velocity v of flow of

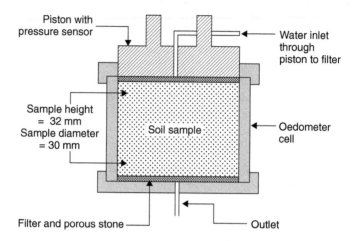

Figure 5.2 Oedometer for permeation of soil samples and recording of swelling
        pressure.

the permeating liquid, to the gradient denoted as i. Since tests are conducted
on the bulk material, and since K is computed (not measured) from recordings of flow rate under a specific gradient, no attention is given to the
nature of the material or its constitution. The oedometer set-up shown in
Figure 5.2 is shown in Figure 5.3 under test conditions where the permeating fluid is applied under a specified constant head to the test sample. The
quantity of fluid Q flowing through a unit time period is collected at the
outlet end in a calibrated flask. Since the cross-sectional area A of the sample is known, the permeating velocity $v = Q/A$ for the unit time period, can
be related to the test-specified gradient i via K as given in the Darcy model:
$v = Q/A = Ki$.

A necessary limiting condition for constant head permeation, in the test
set-up shown in Figure 5.3, is the gradient applied. The gradients used
should impose movement of the permeating fluid without inducing local
piping and erosion or consolidation. The question of limiting the hydraulic
gradient used for tests to provide experimental information to allow for
computation of the hydraulic conductivity coefficient K takes on greater significance when the internal forces developed in swelling clays are called into
play. For swelling soils, computations to arrive at the K value need a proper
consideration of the internal constitution and internal forces. In the standard
procedures, the soil sample in the test cell is assumed to be a homogeneous
porous medium. The method for determination of the conductivity shows no
direct concern for the state of the material, and particularly for the internal
diffuse double-layer (DDL) forces associated with swelling clays. In practice,
a water pressure is usually applied also on the outlet end to guarantee water
saturation. This is not necessary in testing dense smectite clays.

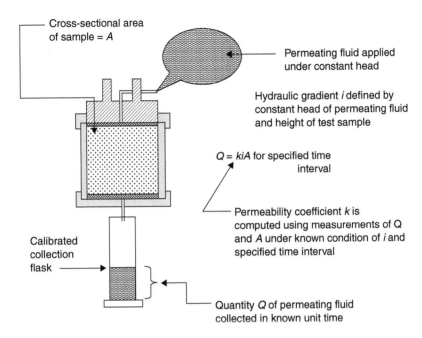

Cross-sectional area
of sample = A

Permeating fluid applied
under constant head

Hydraulic gradient *i* defined by
constant head of permeating fluid
and height of test sample

$Q = kiA$ for specified time
interval

Permeability coefficient *k* is
computed using measurements of Q
and A under known condition of *i* and
specified time interval

Calibrated
collection
flask

Quantity *Q* of permeating fluid
collected in known unit time

Figure 5.3 Illustration of constant head hydraulic conductivity testing to obtain test
values of Q under specified constant head and time interval.

In Section 5.2.1, we have discussed the relative mobility of the water in
wider voids pores and the water associated with small voids. In particular,
the hydrates occupy a special place in the determination of the hydraulic
conductivity of swelling soils. To highlight the significance of these, we can
conduct a simple experiment where the hydraulic gradient *i* can vary from
negative to positive values. Figure 5.4 shows the use of a double Mariotte
tube set-up with the oedometer cell as the reservoir that delivers the permeating fluid. As is well known, since the tops of the tubes are sealed,
the reservoir head is established at the elbow junction air entry shown in
Figure 5.4. By this means, one is able to deliver the volume of fluid in
the reservoir tubes at a constant head. Furthermore, by moving the total
double Mariotte tube up or downwards one can change the hydraulic head
for different sets of tests – to determine the influence of the hydraulic head
on the outflow characteristics.

In the set-up shown in Figure 5.4, water entry into the sample occurs in
response to the osmotic potential in the soil. The roles of the soil
microstructure and the DDL forces are best specified in terms of the matric
ψ_m and osmotic ψ_π potentials. Water drawn into the sample because of the
DDL forces will not be released. Outflow will only occur when the water-
holding forces associated with the DDL are overpowered. This will occur

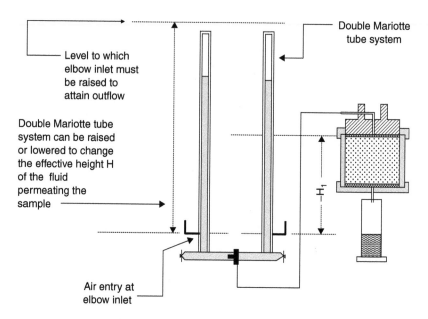

Figure 5.4 Water entry experiment with negative hydraulic head for permeating
fluid (-H1) on swelling soil in oedometer cell. Water entry into sample is
noted by reduced levels of fluid in the Mariotte tubes. No outflow occurs
until the level of the elbow inlet is raised to some point where H reaches
a high positive value. The experiment demonstrates the role of
microstructure and DDL forces in the soil.

when the double Mariotte tube system is raised to a level where the
elbow inlet height can establish a hydraulic head H that will overpower the
water-holding forces in the soil. At this stage, a critical gradient i can be
specified (Figure 5.5). Section 8.3.7 discusses this phenomenon further.

Figure 5.5 shows the relationship between outflow Q and the gradient i
applied, for tests on low-to-moderate swelling soils. For smectites, the separation distance between the apparent threshold gradient i_0 and the critical
gradient i_c is reduced to a very low value. Furthermore, the magnitude of the
critical gradient i_c will be larger than that established by low- and moderate-
swelling clays. The direct proportionality between flow rate and hydraulic
gradient, required in the Darcy model, is not obtained when the hydraulic
gradient is below the critical gradient i_c as shown in Figure 5.5. One might
ask, at this point, whether initial water uptake in response to the microstructure and DDL forces (internal gradients) is accommodated in the Darcy
model. Calculations or measurements are needed to establish the magnitude
of the pseudo hydraulic gradients obtained under the internal gradients.
These need to be incorporated with the externally applied positive gradients
for proper calculation of the effective Darcy permeability coefficient.

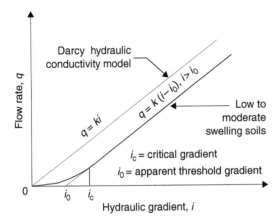

Figure 5.5 Schematic diagram showing critical and apparent threshold gradients obtained in hydraulic conductivity tests on low to moderate swelling soils (adapted from [1]).

5.2.3 Microstructural parameters related to the hydraulic performance of clays

Microstructural parameters introduced some decades ago, that is, the size (maximum diameter) of discernible voids and the P/T ratio with P representing the sectioned voids and T the total section area [2], can be related to the bulk hydraulic conductivity of natural illitic clays. Figure 5.6 shows good correlation between P/T, which is equivalent to the parameter F_2 derived in Chapter 3, and conductivity. The results shown in the figure also demonstrate the difference between fresh-water and marine Quaternary clays.

The impact of compression of the voids in clay is obvious – as explained in Chapters 3 and 4. For all types of clays, an increase in density will result in a reduction in the amount of larger voids, the outcome of which is an apparently more homogeneous macrostructure. This is the result of a tighter packing of the various individual particles and MUs. For smectite clay, this is well demonstrated by results that show small differences between bulk and gel densities of strongly compressed bentonite. This is in contrast to the nature of the clay where the results indicate that when densities are lower than 1500–1600 kg/m^3, microstructural heterogeneity of the material is evidenced by the significant difference between bulk and gel densities – as indicated in Table 5.1.

The differences observed in hydraulic conductivity between clays with salt and salt-free porewater testify to the importance of the DDL forces and microstructural control. In the normal situation of low-electrolyte concentration in the porewater, the pore voids and channels are filled with soft

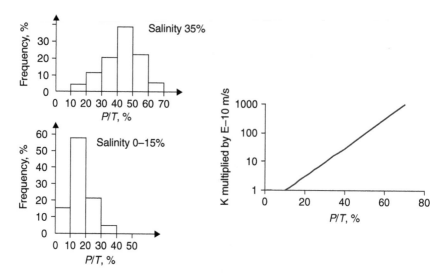

Figure 5.6 Relation between the parameter *P/T* and the hydraulic conductivity evaluated from oedometer tests of undisturbed illitic clay. The wider voids in the marine clays compared to equally dense fresh-water clay explain the higher conductivity of the first-mentioned.

Table 5.1 Bulk and gel densities for MX-80 [3]

Bulk density, kg/m³	Gel density, kg/m³
2130	2000
1850	1650
1570	1150

clay gels. However, when the electrolyte concentration is increased, as for example due to permeation by external sources of fluid with high salt concentrations, gel collapse occurs. This results in the formation of isolated coagulated aggregates. The outcome of coagulated aggregate formation is felt in terms of both partially filled voids and voids empty of clay gels.

5.2.4   Hydraulic conductivity of artificially prepared smectite clay

The F-parameters

We can demonstrate the relevance and usefulness of the procedure for quantifying microstructure described previously in the preceding chapters, by comparing experimentally obtained bulk hydraulic conductivity data for Na and Ca-smectite clay – with calculations using the microstructural parameters F_2 and ρ_{gel}. For the calculations, the basic flow theory assumes the

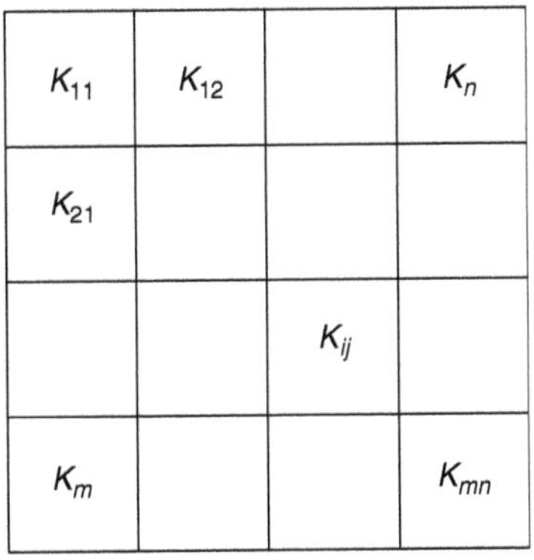

K_{11}	K_{12}		K_n
K_{21}			
		K_{ij}	
K_m			K_{mn}

Figure 5.7 System of elements with different hydraulic conductivity permeated in the horizontal direction.

microstructure to consist of a system of elements with different hydraulic conductivities (Figure 5.7).

The hydraulic conductivity K of soil with elements of different conductivity can be expressed as follows:

$$K = \frac{1}{m}\left[\sum_{i=1}^{m} n\left(\sum_{j=1}^{n}\frac{1}{k_{ij}}\right)^{-1}\right] \tag{5.1}$$

where K is the average hydraulic conductivity, n is the number of elements normal to flow direction, m is the number of elements in flow direction and k_{ij} is the hydraulic conductivity of respective element.

The cross-sectional areas of individual voids identified in micrographs, together with the corresponding K-value calculated from Eq. (5.1) have been compared with experimentally obtained values. These show very good agreement between the two. Considering the uncertainty in estimating interconnectivity and tortuosity of the gel-filled voids and channels distinguishing only between permeable and impermeable fractions of the sections as a first order, simplification appears to be reasonable. The fraction F_2 representing the permeable gel-filled fraction of representative elementary volume (REV) sections can then be used for calculating K. The conductivity of the clay gels is assumed to be similar to the experimentally determined conductivity of clay with the same bulk density. F-data together with the calculated and K-values are shown in Tables 5.2 and 5.3 for three representative bulk densities. Good agreement is obtained between the model-derived data for artificially prepared Wyoming clay (MX-80) in Na form and experimentally determined results.

The microstructural heterogeneity of the artificially prepared MX-80 clay is a significant factor in the hydraulic conductivity of the clay. In comparison

Table 5.2 Microstructural data and conductivities for MX-80 in Na form (percolation with distilled water [3])

Bulk density, kg/m^3	F_2	Gel density, kg/m^3	Gel con-ductivity, m/s	Calculated bulk cond., m/s	Experimental bulk cond., m/s
2130 Na	0.17	2000	7E–14	E–14	2E–14
1850 Na	0.24	1650	2E–12	4E–13	3E–13
1570 Na	0.40	1150	2E–10	8E–11	8E–11

Table 5.3 Microstructural data and conductivities for MX-80 in Ca form (percolation with strongly brackish Ca-dominated water [3])

Bulk density, kg/m^3	F_2	Gel density, kg/m^3	Gel con-ductivity, m/s	Calculated bulk cond., m/s	Experimental bulk cond., m/s
2130 Na	0.17	2000	2E–13	3E–14	3E–14
1850 Na	0.24	1650	8E–11	2E–12	2E–12
1570 Na	0.40	1150	7E–5	3E–06	2E–09

with natural sedimentary clays of the same density and smectite content, the results indicate that higher hydraulic conductivity values are obtained in the artificially prepared MX-80 clay.

Influence of cation exchange

From the perspective of DDL interactions within the lamellae, cation replacement of Na^+ by Ca^{2+} will raise the hydraulic conductivity. If one assumes the same F_2 values and gel densities as for MX-80 in Na form, and if typical gel hydraulic conductivities for MX-80 in Ca form are used, one obtains the results shown in Table 5.3. In the table, the original Na-MX-80 bulk density is given in the first column. The results shown in other columns refer to the MX-80 in the Ca form. As can be seen, good accord is obtained between calculated and true conductivity data for bulk densities as low as about 1800 kg/m^3.

Erosion

Table 5.3 shows that for the bulk density of 1570 kg/m^3, the calculated values grossly over predict the hydraulic conductivity for the Ca replaced MX-80. At low bulk densities, microstructural heterogeneity, dislodgement of particles and non-homogeneous gel formation in the voids for the Ca form of MX-80, combine to produce situations where void-plugging can severely affect conductivity. This is well demonstrated by the results portrayed in Figure 5.8 for tests on MX-80 at a medium density of

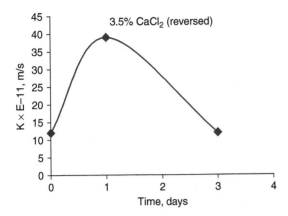

Figure 5.8 Change in hydraulic conductivity of a smectitic clay with a density
of 1800 kg/m³, following flow reversal with 3.5% CaCl₂ solution
under a hydraulic gradient of 30. The conductivity initially
increased in conjunction with disruption of soft clay gels, and
subsequently dropped because of accumulation of released particles
at channel constrictions.

1800 kg/m³ where the permeating fluid consisted of a 3.5% CaCl₂ solution.
The initial rise in the hydraulic conductivity is due to dislodgement of
particles. However, these dislodged particles, together with the resultant
non-homogeneous gel formation in the voids, will become void pluggers.
When this occurs, deterioration in the hydraulic conductivity occurs, as
witnessed by the drop in the hydraulic conductivity after one day [4].

The detailed mechanisms of internal erosion of clay, that is, particle
dislodgement, which is of particular importance for the maintenance of
effective performance of top liners of landfills, brings up the matter of particle bond strength discussed in Chapter 3. High flow rates can cause local
erosion of fully water-saturated smectite clay. The perspective gained from
a study of physical models and experimental results show that a critical rate
of: (a) E–3 m/s is sufficient to produce dislodgement of particles with a size
of 0.5 μm, (b) E–4 m/s for 1 μm particles, (c) E–5 m/s for 10 μm particle
aggregates and (d) E–7 for aggregates – the lower sensitivity for small particles being caused by higher numbers of interparticle bonds [5]. The flow
pattern in a water saturated element of smectite clay with a density of 1600
kg/m³ and containing voids is illustrated by the finite element (FE) calculation shown in Figure 5.9. The vector flow pattern was calculated for the
gradient 1 (difference in metre water head per metre flow length). One concludes that the flow rate in the large central void is in the order of 2E–6 m/s
for a gradient of 100, which is a common value in laboratory experiments.
The disruption of soft clay gels and migration of particle aggregates may
well take place. Verification of this can be seen in Figure 5.10.

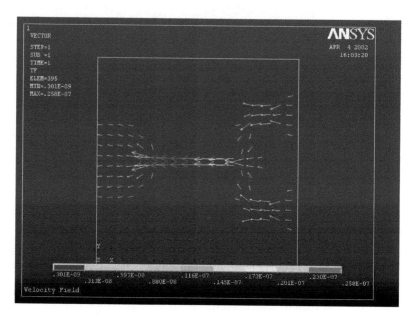

Figure 5.9 FEM calculation in 2D of flow through 30 × 30 μm² clay element of smectitic clay with 1600 kg/m³ density. For a hydraulic gradient of 100 the maximum flow rate represented by the largest vectors is 2E–6 m/s [5] (see Colour Plate IV).

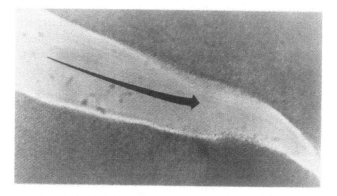

Figure 5.10 Piping in the form of a hydraulic wedge penetrating into soft smectite clay matrix. 20–50 μm aggregates are moved by the flow (E–4 m/s) [6].

Impact of porewater electrolytes on the hydraulic conductivity

The species of electrolytes and their concentrations in the porewater are factors in fluid transport in the clay soil. This is particularly true for clay soils with low-to-medium bulk densities. A lesser influence is felt for soils with much higher bulk densities. Taking the Ca^{2+} ion as an example, where the ion occupies exchange positions and is also present in the porewater, experimental test results show that the calculated and experimental conductivities for a dense smectite clay with a bulk density of 1850 kg/m^3 permeated with sea water were both about 2E–11 m/s. This conductivity value is higher than that obtained for distilled and Ca-rich water because of stronger coagulation and more erosion-resistant gels. For a medium dense smectite clay with a bulk density of 1570 kg/m^3, the theoretically computed conductivity is overpredicted.

Impact of saturation with organic ions on the hydraulic conductivity

We will take the HDPy$^+$-treated clay described in Chapter 4 as an example of the behaviour of a smectite clay saturated with organic cations and molecules. The hydraulic conductivity of the clay is specified in Table 5.4. For a density of 1630 kg/m^3, the conductivity is found to be about 50 times higher than that of untreated MX-80, and about twice as high for the higher density of 1760 kg/m^3. This is explained by the difference in F_2, that is, 0.51 of the HDPy$^+$-treated clay due to the presence of channel systems, and about 0.30 for the untreated clay.

Impact of tortuosity

To account for the effects of tortuosity on hydraulic conductivity, one can use analytical-computer models to investigate various scenarios of path tortuosity. Experiments on replicate samples cannot provide the necessary information because of uncertainties or technical inadequacies in physical determination of the actual flow paths and configurations. One needs to rely on conceptual flow models with computations in 3D. To illustrate the impact of tortuosity on hydraulic conductivity, we will use the 3D flow models with code 3Dchan, developed by Neretnieks and Moreno [7] for our discussions.

The microstructure of the clay considered for analysis is assumed to have an orthogonal pattern of interconnected channels filled with permeable clay

Table 5.4 Bulk hydraulic conductivity (K) of untreated and HDPy$^+$-treated MX-80 clay saturated and permeated with distilled water [3]

Density at saturation, kg/m^3	K, m/s of untreated MX-80	K, m/s of HDPy$^+$-treated MX-80	F_2
1630	2E–12	E–10	—
1760	5E–12	E–11	0.51

gels. It is believed that this configuration is a reasonable representation of the general microstructural heterogeneity of the type of clay under study. Estimation of the size and frequency of the channels containing clay gels can be made along the following guidelines [8]:

- The channels have a circular cross-section.
- The diameter of the widest channel is: (a) 50 μm in smectite (MX-80) clay with a bulk density of 1570 kg/m³, (b) 20 μm in clay with 1850 kg/m³ density and (c) 5 μm in clay with 2130 kg/m³ density, as estimated from TEM. The size of the channels can be taken to have a normal statistical distribution (Figure 5.11).

Using these guidelines, the size distributions in Table 5.5 are obtained. Gel-filled voids with a diameter less than 0.1 μm are considered to be of no importance to the bulk conductivity.

The conceptual model and the computational code consider that to all intents and purposes, all permeating water will flow in the three-dimensional network of gel-filled channels (Figure 5.12). The shape of

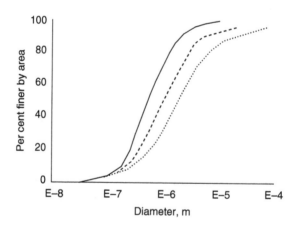

Figure 5.11 Assumed size distribution of pore size in clays A (full line), B (broken line) and C (dotted line) [8].

Table 5.5 Number of differently sized channels per 250 × 250 μm² cross section area representing REV

Bulk density kg/m³	Number of 20–50 μm ch.	Number of 5–20 μm ch.	Number of 1–5 μm ch.
2130	0	0	135
1850	0	10	385
1570	2	85	950

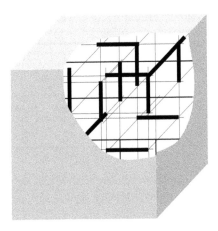

Figure 5.12 Schematic view of the 3D conceptual model (see Colour Plate XI).

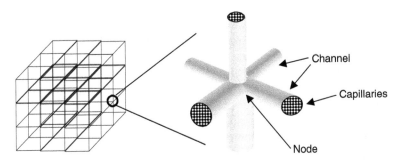

Figure 5.13 Channel network mapped as a cubic grid with gel-filled channels intersecting at a node in the grid.

the channels is characterized by their lengths, widths, apertures and transmissivities, which are all stochastic. The rest of the clay matrix is assumed to be porous but largely impermeable.

Calculation of the bulk hydraulic conductivity can be made by assuming that a certain number of channels (commonly 6) intersect at each node of the orthogonal network. Each channel in the network consists of a bundle of N capillaries with circular cross-section. Figure 5.13 shows a small part of the channel network and illustrates that the channels have different diameters.

The numbers of channels that are assumed to have the length L, contain bundles of N capillaries with a diameter (d). These must be proportional to the channel width and should match the total porosity of the clay. After complete grain expansion, the voids filled with homogeneous clay gels are assumed to have a normal size distribution with the same intervals as in the

Table 5.6 Hydraulic conductivity (K) of three clay types prepared by compacting MX-80 powder and saturating them with electrolyte-poor water (n is the porosity)

Bulk density of air-dry powder kg/m^3	Dry density, kg/m^3	Density at saturation, kg/m^3	n	Calculated K, m/s	Experimental K, m/s
2000	1800	2130	0.13	3E–12	2E–14
1500	1350	1850	0.20	1.3E–11	3E–12
1000	900	1570	0.47	2.4E–10	8E–11

2D model, that is, 1–5 μm for the clay with 2130 kg/m³ density, 1–20 μm for the clay with 1850 kg/m³ density and 1–50 μm for the clay with 1570 kg/m³ density.

The computer code generates a certain number of channels for a given volume [7]. Using the Hagen–Poiseuille law, the flow rate through the channel network is calculated for given boundary conditions assuming a pressure difference on the opposite sides of the cubic grid, and no flow across the other four sides. Table 5.6 specifies the density, total porosity n and resulting bulk hydraulic conductivity of the considered clay types.

From the results shown, one finds that the calculated bulk conductivity of MX-80 clay with the density 1570 kg/m³ at saturation is in the same order of magnitude as typical experimental data. However, the computational model significantly over-predicts the conductivity of the denser clays. A major reason for this is because the model does not take into account the fact that as the density becomes higher, there will be variations in cross-sections between channels, and also along the paths of individual channels, that is, along L. The assumption of a constant cross-section along length L is not totally valid. Thus, the narrowest parts, that is, the smallest cross-sections, will control the transport of the fluid. We note that since the conductivity of a capillary is an exponential function of the diameter, the net effect of constrictions imposed by the smaller cross-sections is more obvious for dense than for soft clays.

Experience from early modelling studies on microstructural constrictions [9] suggest that the matter of constrictions can be accounted for by assuming that about 10% of the length of all channel capillaries have a radius that is 2/3 of the diameter of the remaining length. The impact of such an assumption on the conductivity can be readily determined using a theory proposed for capillaries with two dimensions. Taking the large diameter as d_1, and the small one as d_2, and $d_1 = 4d_2$, and the length of the small diameter as L_2, and also $L_2 = 0.1(L_1 + L_2)$, one gets the equivalent diameter as $d_2/(L_2)^{1/4}$. This will give a reduction of the K-values according to Table 5.7. The corrected theoretical conductivity values agree well with the experimental data except for the densest clay. For the densest clays, some further reduction in capillary diameter is required.

Table 5.7 Comparison of corrected theoretical hydraulic conductivity and experimental values

Density at water saturation, kg/m^3	Corrected theoretical K, m/s	Experimental K, m/s
2130	E–13	2E–14
1850	5E–12	3E–12
1570	6E–11	8E–11

The 3D computational model confirms that the concept of flow channels filled with gels containing numerous capillaries is a reasonably relevant theoretical model. It is important to remember that the true void arrangement and the gels and constrictions, etc. are stochastic in nature, with large variations in particle spacing and interaction that are not included in the 3D flow model.

5.2.5 Hydraulic conductivity of artificially prepared non-smectite clay

General

The fact that smectite has the lowest hydraulic and gas conductivities of all clay types considered in this book is obvious from the results shown in Figures 5.14 and 5.15. They compare the conductivities of montmorillonite (smectite), kaolinite and illite clays obtained from oedometer tests using pure minerals of montmorillonite, illite and kaolinite. This explains why the conductivity data for montmorillonite are lower in comparison to commercial montmorillonite-rich clay like the MX-80. From Figure 5.14, we note that smectite and kaolinite represent two ends of the conductivity spectrum, with smectite showing conductivity values that are four orders of magnitude lower than the kaolinite. The kaolinite shows performance characteristics not unlike that of very fine-grained silt. Illite (Figure 5.15) shows conductivity values that are intermediate to the smectite and kaolinite clays.

Clays with an appreciable amount of mixed-layer minerals with smectite as one component will generally have sufficiently low hydraulic conductivities to be useful for use as barriers and liners for isolation of solid waste. A good example of this is the German Friedland Ton, termed FIM in this book. Its conductivity is shown in Figure 5.16 as a function of density and porewater salinity. The performance of this clay is relatively insensitive to a high salt content in the porewater – as explained by the low gel-forming capacity. The powder grains of this dried clay expand in a manner similar to smectite-rich clays. However, the voids between the grains are not completely filled with clay gels. The microstructural parameter F_2 is higher than that of a correspondingly dense MX-80 clay. For a density of 2000 kg/m^3

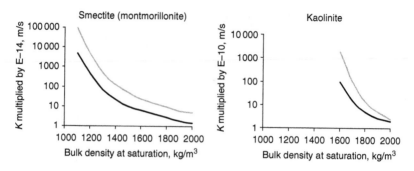

Figure 5.14 Hydraulic conductivity of clays consisting of pure smectite (montmorillonite), kaolinite and illite. Upper curves represent saturation with seawater.

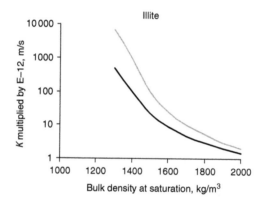

Figure 5.15 Hydraulic conductivity of nearly pure illite. Upper curve represents saturation with seawater.

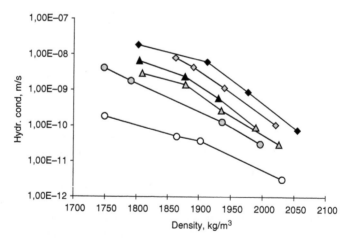

Figure 5.16 Impact of salt water on the hydraulic conductivity of FIM clay. From above: 20% NaCl, 20% $CaCl_2$, 10% NaCl, 10% $CaCl_2$, 3.5% $CaCl_2$, distilled water.

and saturation with low-electrolyte water, the hydraulic conductivity of the latter (MX-80) is about 100 times lower. However, in the presence of a strong $CaCl_2$ solution, the difference is significantly smaller.

5.3   Gas conductivity of artificially prepared smectite clay

5.3.1   *General*

The processes and mechanics of gas permeation in and through a clay soil are not fully appreciated or understood. Nevertheless, the performance of a clay in respect to gas permeation is of considerable practical importance for: (a) clay-sealing of underground facilities and (b) understanding gas prospecting and exploitation in areas such as the North Sea. Recent data from Russian gas and oil fields show that sediments consisting of clays dominated by kaolinite, illite and chlorite are low-permeable to permeable (E-8–E-7 m/s) to gas and oil. On the other hand, mixed-layer smectite-illite exhibit apparently higher permeability characteristics, with values for gas and oil diffusion in the order of $D = $ E-4–E-2 m²/s. Only montmorillonite clays with porosities lower than 30% and a Na/Ca ratio of 4–12 are to all intents and purposes practically tight [10].

In practice one can assume that the gas conductivity is about a thousand times higher than that of water. This means that once gas has made its way through a clay buffer and through the surrounding host material such as the more permeable geological units, the gas will flow at a rate that is controlled only by the available gas pressure – as opposed to conductivity of the host material. The most important issue is therefore the threshold gas pressure that initiates gas penetration through the buffer clay, that is, the *'critical gas pressure'*. For temporarily stagnant gas in clay voids, the solubility of the gas in question – air, hydrogen or organic gases – is an important matter. When dissolution occurs, the bubbles will shrink and the dissolved gas will diffuse out of the system. In smectite-rich clay, this is associated with microstructural self-healing, that is, convergence of widened voids and subsequent growth of clay gels by stacks of lamellae released from the denser matrix.

Microstructural heterogeneity has a decisive influence on the critical gas pressure. This is because gas migrates along paths of least resistance – that is, along continuous channels of neighbouring larger gel-filled voids where the capillary retention is at minimum, or when the bond strength of adjacent particles is most easily overcome [11, 12]. The resistance to gas-induced displacement of the clay matrix in dense smectite clay is directly related to its swelling pressure since gas penetration requires separation of dense matrix components ('fracturing'). For very dense smectite clays, the critical gas pressure should in fact be close to the bulk swelling pressure, which is confirmed by the results shown in Table 5.8.

For lower density smectites, we see from Table 5.8 that the critical gas pressure is again close to the swelling pressure of the material – except that in this case, the swelling pressure of the lower density smectites is one to

Table 5.8 Experimentally determined critical gas pressure in
MPa for MX-80 in Na form [3]

Density at saturation, kg/m³	Experimentally determined critical gas pressure, MPa	Swelling pressure of dense clay matrix in Na MX-80
2130	20	20
1850	2	1
1570	0.1	0.2

two orders of magnitude smaller than for the dense smectite. This is to be expected since the interparticle spacings are larger. For equilibrium to be maintained, the pressures in the gels in the voids must balance the swelling pressure. Gas migration in these voids will be through the gels, and since the gel strengths are entirely reflective of the swelling pressures, it is expected that the critical gas pressure will be close to the swelling pressure of the clay. Figures 5.17 and 5.18 show the recorded gas flow in tests with clay soils prepared by compaction in a 'megapermeameter' with 780 mm inner diameter and 300 mm height after saturation with artificial seawater through filters at the base and top. A backpressure of a few hundred kPa was applied to maintain a high degree of water saturation.

Tests on breakthrough pressures for two types of clays show some very interesting performance characteristics. One of the clays was commercial clay powder, rich in montmorillonite (GEKO/QI, Sued-Chemie) compacted to a dry density of 1080 kg/m³ (1680 kg/m³ at saturation) with the hydraulic conductivity and swelling pressure of 4E–11 m/s and 200 kPa, respectively. The other was a mixture of 10% fine-grained clay of the same type and 90% graded granitic ballast compacted in thin layers to a dry density of 1920 kg/m³ (2200 kg/m³ at saturation). The hydraulic conductivity was 2E–9 m/s for a hydraulic gradient of 30 metre per metre) and the swelling pressure about 50 kPa. For the pure clay, breakthrough occurred at a gas overpressure of 190 kPa about 5 days after reaching the critical pressure. For the 10/90 mixture this pressure was only 15 kPa and breakthrough occurred after about 10 minutes. This low pressure is consistent with the properties associated with the low density of the clay gels in the voids between the ballast grains.

A slow build-up of the gas pressure causes partial consolidation of the softest clay gels. When the critical pressure is reached, gas movement becomes faster. This will result in piping. The process of gel healing and recovery is not sufficiently fast to prevent gas penetration. From experimental results, it is observed that porewater movement resulted from gas pressurization, and that a discharge of about 0.02–0.2% of the water content was obtained before evidence of gas penetration was recorded. More evidence is needed to determine whether the critical breakthrough pressure is lower or higher for a very slow gas pressure increase.

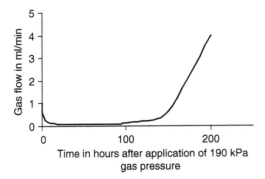

Figure 5.17 Example of the gas penetration process at the critical pressure for a bentonite clay with 1680 kg/m³ density saturated with seawater in a megapermeameter test. The initial drop of inflow rate was caused by the elastic strain of the permeameter and clay. The increased inflow rate after about 80 hours was due to the successively increased pressure gradient when the gas front approached the outlet. After 200 hours, steady flow was approached.

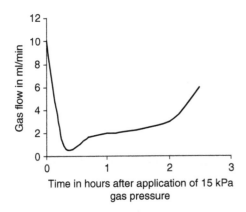

Figure 5.18 Example of the gas penetration process at the critical pressure for 10/90 mixture of Na bentonite and graded ballast with a density of 2200 kg/m³ and saturated with seawater.

5.3.2 Applicability of microstructural parameters

As with hydraulic conductivity, gas penetrability can be estimated using microstructural parameters. In spite of the relatively large variations in width and linearity of the channels and density of their gel fillings in the particle network, the same F_2-value representing the permeated part of any cross-section of a REV can be used. This parameter is hence related not only to porewater flow but also to gas migration. Referring to Table 5.2, F_2 would be about 0.30 for the bentonite clay used for the test reported in Figure 5.17 whereas for the 10/90 mixture (Figure 5.18), one expects F_2 of the clay gel in the voids to be about 0.4. However, this figure should be somewhat lower because part of a cross-section through the latter soil is occupied by solid impermeable grains. This means that gas conductivity for the two soils should not be very different. This is in fact proven by the graphs in Figures 5.17 and 5.18.

The most important property, the breakthrough pressure, is more related to the pressure conditions and hence to the microstructural parameter F_3. We will show later in this chapter how the swelling pressure depends on this parameter. For this section, we will confine ourselves to a discussion of the impact of gel density in the two gas tests previously described. For the gas-tested Na bentonite with a density of 1680 kg/m^3 after saturation with sea-water, this density was around 1500 kg/m^3. In the case of the 10/90 mixture, the density was not higher than about 1300 kg/m^3. The difference in swelling pressure, and therefore critical breakthrough pressure, is simply and fully explained by the different clay gel densities.

5.4 Ion diffusion

5.4.1 General

Clay soil properties and characteristics affecting diffusive transport processes of ions in engineered clay liners and buffers are of fundamental importance in the selection of clay materials for hazardous waste isolation barriers and liners. The problem in predicting and evaluating ion migration in clays in general, and smectite clays in particular, is that ion transport processes in a clay-water system involve a host of chemical reactions that depend on several factors. Paramount amongst the factors controlling the transport and fate of contaminant ions are: (a) type and activity of soil particles, (b) cation-exchange capacity (CEC) and specific surface area (SSA) of the soil solids, (c) chemistry of the porewater, (d) nature, species and concentrations of the contaminant ions or chemicals, (e) redox potential and pH and (f) kinds of soil microorganisms in the system. For this section, we will consider, in a general manner, how diffusive migration depends on the microstructure of the soil in the engineered barrier system.

5.4.2 Ion flux capacity

Everything else being equal, the transport rate of ion species in clays depends on their diffusivity and their concentration gradients. The diffusion transport capacity expressed by the 'effective' diffusion coefficient takes into account the actual 'effective' porosity existent at the microstructural level. This is in contrast to the 'apparent' diffusion coefficient, which is a general measure of diffusion derived directly from recorded concentration profiles of the bulk sample. Cation diffusion occurs in several ways, that is, in continuous water-filled voids, along particle surfaces with electrical double-layers, and through the interlamellar space in smectites. The latter two mechanisms involve ion-exchange mechanisms for which the sorption parameter, K_d, is used. In practice, the ion transport capacity can be predicted by applying Fick's law and relevant values of the coefficient of the density-related 'effective' diffusion, D_e. The density of the clay plays an important role – especially for anions as illustrated by Figure 5.19.

The diffusive anion transport capacity is proportional to the ratio of the pore space of the voids between the stacks of smectite lamellae since they are excluded by Donnan effect from the interlamellar space. With increasing density there is a strong reduction of the available space for migration, and the diffusion coefficient of anions therefore drops significantly. Since many cations move both by pore diffusion and surface diffusion, the retarding effect resulting from increased densities on the diffusion capacity of cations is limited, especially for monovalent ions.

Since the F_2 parameter represents the part of the clay in which pore diffusion takes place, it determines the diffusive capacity of certain ions and molecules. Thus, the ratio of the void volume available for uniaxial diffusion and the total volume should be approximately proportional to F_2 for anions, and to unity for lithium and sodium. The ratio of the ion diffusion capacity for iodine and sodium, for example, should be about 0.17 for a bulk density of 2000 kg/m^3. This is about 4 times higher than the experimentally determined ratio [3, 14]. The main reason for this discrepancy is that channel tortuosity and constrictions have not been accounted for. When we apply the correction of F_2 used in calculating the hydraulic conductivity caused by constrictions, we will obtain a ratio for iodine and sodium diffusion capacity that is very close to recorded values.

Since F_2 is density-related it is expected that prediction of ion fluxes using this parameter would show a clear drop in anion flux with increasing

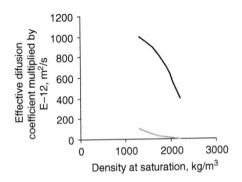

Figure 5.19 Measured effective diffusivities for smectite clay [13]. Upper: monovalent cations, lower: monovalent anions such as chlorine.

clay density. This is due to the fact that anions are prevented from entering interlamellar space by Donnan exclusion. Cations on the other hand would not be affected as much by an increase in clay density. This is demonstrated by the results shown in Figure 5.19. Although it would be tempting to conclude that one should strive to obtain very high clay buffer densities, there is a price to pay somewhere along the line. As we have discussed previously, very high densities of smectite will also mean that very high swelling pressures can be generated. These can cause upheaval of landfills if smectite-rich bottom liners are compacted too strongly (i.e. too densely). In repositories, it is possible for these very high swelling pressures to cause failure in the rock surrounding the high level radioactive wastes (HLW) canisters.

5.5   Expandability

5.5.1   *General*

Most clays show some expandability (swelling) upon water uptake, and thereby some self-healing ability, that is, ability to heal intrinsic macroscopic cracks and apertures. The large SSA, in combination with the properties of the lamellae, give smectites high expansion (swelling) capability and therefore high potential for self-healing upon water uptake. As we have stated previously, this attribute is of considerable importance in preserving the integrity of clay-engineered barrier and liner systems. Since cracks, openings, channelization and local shrinkage are some of the negative consequences that can arise because of gas migration, oxidation–reduction reactions and moisture depletion, to name a few, the ability of the material to maintain internal integrity through self-healing is of utmost importance. Smectites have the potential to perform well in this respect.

5.5.2   *Swelling pressure*

The swelling pressure exerted on the physical boundaries of smectite clay seals is caused by a combination of: (a) the true 'disjoining' pressure caused by the interlamellar water films which strive to grow to a certain finite thickness if expansion can take place and (b) the osmotic pressure caused by the charge conditions at the outer boundaries of the stacks of lamellae (Chapter 4, Section 4.3.6 of this book). The swelling pressure-time relationship shown in Figure 5.20 has been obtained from swelling pressure tests on montmorillonite-rich clay, using oedometers equipped with stiff pressure cells. The clay sample was prepared by compacting air-dry clay powder directly in the oedometer, and was allowed to sorb water after final sample preparation. It is interesting to follow the progress of a test sample during the uptake process. For a smectite-rich clay with a dry density of at least a few hundred kg/m^3, the air enclosed in the powder becomes pressurized in the hydrating clay. The air dissolves and diffuses out of the system – a process that occurs even if water uptake is allowed

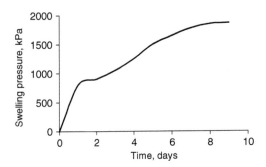

Figure 5.20 Development of swelling pressure of maturing clay with a montmorillonite content of about 70% and a dry density of 1390 kg/m^3 (1875 kg/m^3 at complete saturation). Saturation with distilled water.

to occur from one end only. The first peak reached after the first day is due to evolution of the microstructure. From a physical standpoint, we can describe this as the expansion process of the individual dense clay grains and subsequent movement into more stable positions. This process is by no means complete in the early days because the process of expansion depends on water uptake rate, which is a slow process. This process, which has been discussed previously, is a slow process. In consequence, continuous readjustment of the expanding and expanded grains to more stable positions occurs. Typical smectite swelling pressure data are collected and shown in Table 5.9.

The smectite-rich clays MX-80 and IBECO, which are commercially available in large quantities, have montmorillonite (75–90%) as their major smectite constituent. At low densities, saponite and beidellite behave as montmorillonite. They develop higher swelling pressures at high densities. For densities higher than about 2000 kg/m^3 at fluid saturation, the chemical composition of the porewater does not exercise much influence on the swelling pressure because the DDL forces are practically non-existent. The low amount of water uptake at very high densities is to all intents and purposes pseudo-crystalline interlamellar water. Calculations will show that when the maximum number of hydrates in interlamellar space is achieved, the samples will no longer be classed as very dense samples. Water uptake beyond hydrate water will be by DDL forces, at which time the density of the samples will be considered in the medium density range. At this time, the influence of the cations species becomes felt. For lower densities, and hence lower swelling pressures, the chemical composition of the porewater becomes an important factor – as has been explained in Chapter 4 (cf. Figure 4.15).

In this context it is interesting to note that the previously discussed Friedland clay with its dominant content of mixed-layer smectite/muscovite is less affected by variations in porewater chemistry than the smectite-rich

Table 5.9 Swelling pressure (p_s) in MPa of well-
characterized smectite-rich materials at
saturation with distilled water

Density at saturation, kg/m³	1800	2000
MX-80	0.8–0.9	4–5
IBECO, Na	0.6–1	4–5
IBECO, Ca	0.2	5
Beidellite	1.5	4.2
Saponite	2.5	8.8

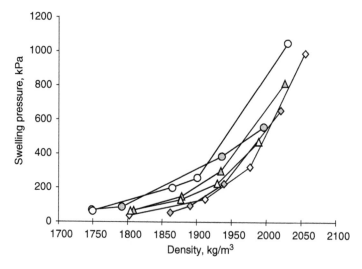

Figure 5.21 Swelling pressure tests on saturated samples of FIM clay. From above:
Distilled, 10% CaCl$_2$, 3.5% CaCl$_2$, 10% NaCl, 20% CaCl$_2$, 20% NaCl.

clays. The pressure is lower at all densities, however, as demonstrated by the
diagram in Figure 5.21. This is attributable to the lower expandability of
the crystal lattice and the lower content of clay gels in the voids between the
dense expanded grains. We should note that for a density, at saturation with
brine of 20% NaCl, of 1900 kg/m³ the clay still develops a swelling pressure of at least 100 kPa. This is considered to be a requirement for
supporting the roof of backfilled tunnels and drifts in HLW repositories.

The F-parameters

Chapter 2 has shown that the swelling pressure in the clay soil is caused by the
disjoining forces in the interlamellar space and double-layer repulsion.

Considering the actual force distribution in the common heterogeneous microstructure, the swelling pressure is seen to be proportional to the product of the true swelling pressure of the pressure-controlling component a and the volume ratio $(a^3 - b^3)/a^3$ in Figure 3.28. This ratio is $(1 - F_3,)$, which gives the volume fraction of this component. For a bulk density of 2130 kg/m^3, ρ_a is 2000 kg/m^3 and the true swelling pressure of this component, calculated using the Yong and Warkentin theory [1], is about 11 MPa, assuming isotropic distribution and orientation of the smectite lamellae. The value of $(1 - F_3)$ is 0.93 and in consequence, the product is about 10 MPa. For a bulk density of 1570 kg/m^3, ρ_a is 1750 kg/m^3 and the true swelling pressure is about 0.3 MPa. The value of $(1 - F_3)$ is 0.75 and the product is about 0.2 MPa. The values are in good agreement with the experimental data shown in Table 5.10.

Influence of cation exchange

Applying the same generalization as has been done for the hydraulic conductivity analysis, that is, assuming that the major microstructural features expressed in terms of F_3 are about the same as for saturation with distilled water, one gets for MX-80 in Ca form, the same theoretical swelling pressure as for MX-80 saturated with distilled water when the bulk density is 1850 kg/m^3 and 2130 kg/m^3, respectively. These pressure values agree well with experimental results. For a density of 1570 kg/m^3 on the other hand, the theoretical value is 0.3 MPa. The typical experimental value is only about 0.02 MPa. The discrepancy is due to insufficient sensitivity of the simple model towards densities representing conditions close to complete expansion of the densest part of the clay matrix (a). Altering the porewater electrolyte of smectite-rich clay with a density of 2130 kg/m^3, from a low salinity solution to sea water, does not produce any observable change in the swelling pressure – as has been documented by numerous experiments. For a density of 1850 kg/m^3, the theoretical value is around 1.0 MPa. A typical experimental value is in the order of 0.5 MPa. For the lowest bulk density of 1570 kg/m^3, the theoretical swelling pressure is about 0.15 MPa. Recorded swelling pressures are only a few tens of kPa. The influence of

Table 5.10 Calculated and experimentally determined swelling pressures (p_s) of MX-80 saturated with distilled water

Bulk density, kg/m^3	$1 - F_3$	Density of massive part, kg/m^3	p_s of massive part, MPa	Calculated bulk p_s, MPa	Experimental bulk p_s, Mpa
2130	0.93	2150	15	14.0	14
1850	0.80	1900	1.5	1.2	1.0
1570	0.75	1750	0.5	0.4	0.3

Table 5.11 Swelling pressure (p_s) of untreated and HDPy$^+$-treated MX-80 clay saturated and permeated with distilled water

Density at saturation, kg/m³	p_s, MPa of untreated MX-80	p_s, MPa of HDPy$^+$-treated MX-80
1630	0.35	0.60
1760	0.60	0.70

increased Na concentration in the porewater is similar to the effect obtained for hydraulic conductivity.

Influence of organic ions on the swelling pressure

The influence and impact on swelling pressure due to the presence of organic species in smectite-rich clay can be seen in Table 5.11. For a density of 1630 kg/m³, the swelling pressure is seen to be almost twice as high as that for untreated MX-80. It is, however, about the same as that shown for the untreated MX-80 for the higher density.

The value 0.63 of the factor $(1-F_3)$ for the HDPy$^+$-treated MX-80 clay with a density of 1760 kg/m³ is somewhat lower than that of untreated MX-80 clay with the same density (0.75–0.80). However, the density of the massive parts of the matrix in the digitalized picture in Figure 4.16 is somewhat higher, that is, about 1800 kg/m³. The theoretical bulk swelling pressure, which is the product of $(1 - F_3)$ and the estimated swelling pressure 1 MPa of the massive part should by all accounts, be of the same order of magnitude as the actually recorded pressure.

The reason for the relatively high swelling pressure of low-density smectite clay saturated with HDPy$^+$ is believed to be due to hydration – as indicated in Figure 5.22, which suggests the same diffusion-controlled uptake of water and development of swelling pressure as in the ordinary MX-80 clay. This means that both water molecules and HDPy$^+$-ions populate interlamellar space. In this context, it is of interest to realize that HDPy-treated MX-80 has a much more heterogeneous microstructural constitution than untreated MX-80, and that its gel-forming potential is significantly lower. The HDPy cation is believed to be adsorbed in the interlamellar space in a 'pillared' fashion, which allows for interlamellar hydration. The stacks of lamellae are rigid and require much higher compaction pressure than 100 MPa to yield bulk densities exceeding about 1800 kg/m³ after water saturation.

One can conclude that a number of comparisons between actually determined swelling pressures and theoretically derived values based on microstructural models of clays with different smectite contents, show good qualitative and quantitative agreement for ordinary inorganic cations. However, organic cations and molecules occupying interlamellar space can produce surprises – as in the case of the HDPy cation. Much work and

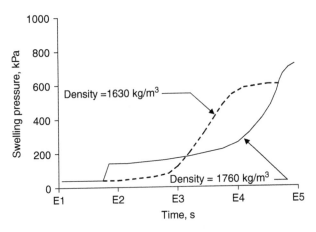

Figure 5.22 Development of swelling pressure in HDPy$^+$-treated MX-80 clay [3].

study is required – especially in respect to the development of simple techniques for quick determination of the smectite content in routine testing of smectitic clays.

5.6   Microbial function

5.6.1   *General*

Interaction of organic species with smectite clays in general has been an important issue for a long time especially with respect to the performance of microbes. In recent years, the matter has become particularly important in conjunction with the design of repositories for containment of radioactive waste. The requirement is that microbes must not penetrate the clay buffers since they can form organic colloids that can migrate to the biosphere. The danger is that these migrating colloids will bring radionuclides with them. The problem has been studied from different viewpoints, one of which is the microstructural constitution of the clay and the possibility for microbes to survive and migrate within this type of system. This subject will be briefly discussed here since the manner of treatment can be used also for such practical situations as top liners of waste landfills.

5.6.2   *Space for microbial survival*

MX-80 clays with densities lower than 1200 kg/m^3 have sufficiently large voids to host large amounts of bacteria and allow them to multiply and migrate by breaking gel bonds (Figure 5.23). However, fixation to clay gels by hydrogen bonds puts limits to their mobility [15]. The frequency of relatively large voids in MX-80 buffer clay with a bulk density of more than

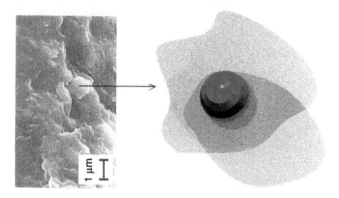

Figure 5.23 Bacterium embedded in montmorillonite clay with a dry density of
1500 kg/m³. Left: SEM picture, right: schematic view of the protruding
bacterium, magnified to some extent (see Colour Plate VII).

1800 kg/m³ may be sufficient to host bacteria. Certain bacteria, such as
Desulfotomaculum nigrificans, which play a role in the degradation of copper canisters, can survive in certain voids in the softest clay gels at this bulk
density – since they provide sufficient access to free water even though
sufficient nutrients for multiplication and spore production may not be
available. At higher bulk densities, bacteria are not able to move in the clay
because of the limited space and the mechanical strength of the clay gel.
They will ultimately lose their potential for producing spores and will die.
This has been validated by recent experiments.

One way of substantiating the geometry and possibilities for survival and
multiplication of microbes such as bacteria and spores is to develop 3D
microstructural models. Assuming microstructural isotropy, transmission
electron microscopy (TEM) pictures can be interpreted in 3D as shown in
Figure 5.24 for MX-80 clay.

5.7   Stress/strain behaviour

5.7.1   *General*

The deformation properties of artificially prepared clays are of great practical
importance: (a) for freshly manufactured big buffer blocks, since they will
be exposed to high stresses in the transport and placement in repositories
for radioactive waste and (b) for top liners of waste landfills upon which
vehicles move in the construction phase. They are even more important for
mature clay buffers because of the requirement to support the heavy weight
imposed by the waste canister resting on the clay buffer, and also because

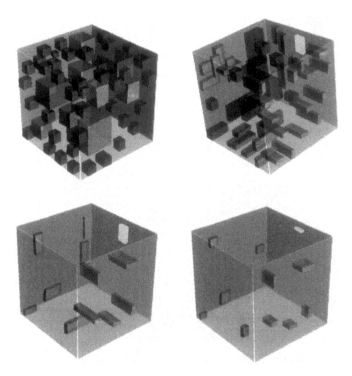

Figure 5.24 3D system of boxes representing voids that are open or filled with soft clay gels in a cubical clay element with 30 μm edge length. Upper left: Bulk density of 1300 kg/m³ with a high frequency of 1–20 μm voids, maximum void size 50 μm. Upper right: Bulk density of 1600 kg/m³ with a moderately high frequency of 1–15 μm voids, maximum size 20 μm. Lower left: Bulk density of 1800 kg/m³ with a low frequency of 1–10 μm, maximum size 15 μm. Lower right: Bulk density of 2000 kg/m³ with very few voids, all smaller than 5 μm (see Colour Plate V).

of the need to sustain tectonically induced shearing of the clay buffer. In this section, we will examine the mechanisms associated with shear straining of matured smectite clay seals (buffers and liners) with special interest on the microstructural performance of the material.

The raw materials used for preparing artificial clay seals are usually bentonites with some slight cementation caused by dissolution/precipitation processes. The cementing substance can be crystalline silicious compounds, amorphous silica, iron compounds or calcite. The relatively strong particle bonds mean that the often smectite-rich beds do not disintegrate spontaneously when wetted by rain or snow (Figures 5.25 and 5.26). However, drying and grinding and ion exchange to Na will increase the activity of these clays to some significant extent.

Figure 5.25 Shaly Canadian bentonite exploited for manufacturing clay powder. The clay is cemented but crushing in the processing plant breaks down the strongest bonds (Photo: M. Gray) (see Colour Plate X).

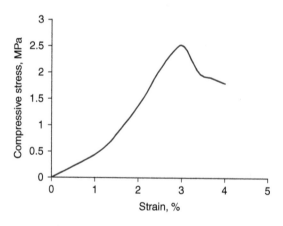

Figure 5.26 Typical brittle-type stress/strain behaviour at uniaxial compression of significantly cemented argillaceous rock. Ordovician illite originating from smectite. Southern Gotland, Sweden.

5.7.2 Shearing mechanisms

The heterogeneity of artificially prepared and matured smectite clay means that the particle bond strength or the energy barrier spectrum is wide, ranging from weak hydrogen bonds in the soft clay gels via shared cationic bonds to strong primary valence bonds in the cementations (Figure 5.27). Applying a shear stress to an element of clay, the weakest bonds are broken first and the stresses transferred to stronger components that deform and yield if the stress is increased. The detailed strain mechanisms illustrated in Figure 5.28, which shows strong aggregates of clay particles and clay gels connecting and surrounding them [16]. The latter yield first, bringing the stronger aggregates in temporary or permanent contact.

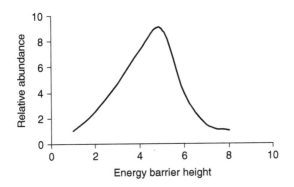

Figure 5.27 Schematic energy barrier spectrum in clay under shearing. The lower levels represented by 1, 2 etc represent weak hydrogen bonds, and points 4, 5 etc. the highest barriers represented by primary valence bonds. At a certain stage and time, the weakest barriers will be broken and the higher ones will be activated. For simple mathematical treatment, the peak shape of the spectrum can be replaced by a box-shaped distribution.

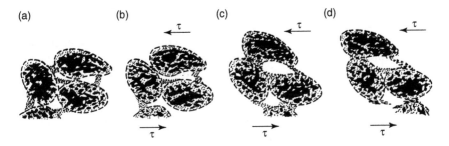

Figure 5.28 Consecutive stages in the evolution of shear strain of microstructural network of clay particles. (a) Before loading, (b) instantaneous shear and formation of slip units due to application of a shear stress τ, (c) formation of slip domains accompanied by healing and breakdown, (d) failure.

5.7.3   *Shear resistance*

General

Conventional thinking imagines that smectite clays follow the same rules as any soil. This implies that the soil obeys the same effective stress concept and the Mohr–Coulomb model for determination of the shear strength of the soil. The two-parameter Mohr–Coulomb model is best applied to a rigid-plastic material that does not undergo volume change during shearing. Applying conventional theories to the shear resistance analysis of smectites poses some very interesting observations – especially when expressing the shear strength in terms of such soil properties as: (a) *cohesion*, using the intercept of the failure curve on the shear resistance axis (ordinate on the Mohr–Coulomb plot) and (b) *friction*, using the angle of friction as the basic expression of the frictional properties of the soil. The applicability of this conventional approach has been well discussed [17], particularly in respect to the situation where rigid grains do not exist, and where frictional contact between the non-rigid grains is not manifested. The classical theories of granular mechanics that apply well for granular soils have severe limitations when applied to clays, and especially and particularly to swelling clays such as smectites. The development of hydrates and the interactions from DDL forces, combined with the basic lamellae structure, make it difficult to apply granular contact theories. In particular, the concept of positive porewater pressure and its measurement need to be critically explored and understood. Laboratory tests to accurately determine shear resistance mechanisms are difficult because of the extremely slow consolidation and expansion rates and associated transient microstructural changes.

The F-parameters

The shear resistance of smectite clays can be conceptually imagined by applying the same sets of consideration as was performed for the swelling pressure. Considering the actual force distribution in the heterogeneous microstructure, the shear resistance should be proportional to the product of the true swelling pressure of the pressure-controlling component a and the volume ratio $(a^3 - b^3)/a^3$ in Figure 3.28. This ratio is $(1 - F_3)$. This establishes the volume fraction of this component. Referring again to the three selected bulk densities of the reference MX-80 clay, one finds that for the bulk density of 2130 kg/m^3, ρ_a is 2000 kg/m^3. If we use the conventional soil shear resistance model, we find that the true shear resistance of this component is the product of the swelling pressure and tan ϕ, where ϕ is the friction angle. From numerous laboratory tests, this friction angle has been determined to be about 10°.

The swelling pressure is 10 MPa, as derived earlier, and the shear resistance hence about 2 MPa. This is almost equal to the shear strength of the soil obtained from triaxial tests. For a bulk density of 1570 kg/m³, ρ_a is 1750 kg/m³ and the true swelling pressure is about 0.3 MPa. The value of $(1 - F_3)$ is 0.75 and the product will be about 0.2 MPa, meaning that the shear strength is about 40 kPa. This is within the same order of magnitude and is consistent with the results obtained from numerous experiments. For lower densities the shear resistance is very low.

5.7.4 Creep

Smectite clays belong to a class of clays that is usually known as *plastic clays* in conventional soil mechanics practice. These kinds of clays are known to undergo significant time-dependent strain (creep) under constant shear stress conditions. Creep is an important phenomenon that determines the rate of settlement of heavy waste canisters: (a) because of their weight and (b) because of changes in stress conditions in the clay embedment of canisters generated by tectonically induced shearing.

The creep of clay at constant deviatoric stresses depends on the stress and the ambient temperature. Thermally activated slip is the most likely rate-determining process [9, 16]. In view of the heterogeneity of the material, from the point of view of local stress and chemical structure, we can consider that at a given point, j in the material, slip will be deterred by an energy barrier denoted as u_j. This deterrent is governed by the intrinsic nature of the obstacle, u_{j0}, as well as by the local deviatoric stress σ_j acting on it at time t. One can therefore represent the stress- and time-dependence of u_j by the expressions:

$$u_j = u_j(u_{j0}, \sigma_j) \tag{5.2}$$

$$\sigma_j = \sigma_j(t) \tag{5.3}$$

Although earlier rate theories were based on the assumption that the intrinsic energy barriers are all of the same type, modern theories that consider the structural heterogeneity take into account the actual distribution as indicated for example in Figure 5.27. This leads to a strain rate for constant stresses as shown in Eq. (5.4). The stresses imposed are sufficiently high to initiate creep, but must not exceed those values that will initiate failure.

$$d\gamma/dt = \beta T \sigma^n (t + t_0) \tag{5.4}$$

where β is the deformation modulus, γ is the shear strain, σ is the shear stress, t is the time after onset of creep, t_0 is the constant and T is the temperature.

Figure 5.29 shows the generalized evolution of creep strain according to the described creep theory. It is seen that the stochastic mechanics-based theory can adequately predict the creep strain. We need to look at the meaning of negative t_0s, which are representative of cemented clays with a

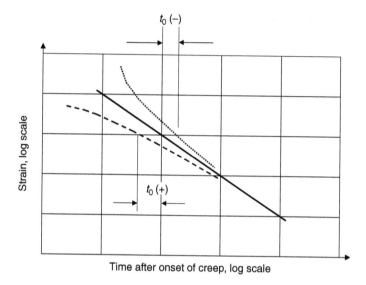

Figure 5.29 General appearance of the creep behaviour of clays. *T* is time after
onset of creep and *dγ/dt*, where γ = shear strain.

supply of very high energy barriers. The negative values mean that the
initial creep rate is high. Positive t_0s represent normally consolidated clays,
including artificially prepared smectite clays. Usually, the straight line for
positive t_0s, imply that the creep rate is proportional to log *t*, and will be
reached after a few days. The shear stress criterion implies, in principle, that
the stress must be in the interval $\tau < \frac{1}{3}\tau_{max} < \frac{2}{3}\tau_{max}$, where τ_{max} is the shear
strength at ordinary quick shear tests [16].

5.7.5   Thixotropy

Smectites are thixotropic. This makes them useful in manufacturing
non-dripping paints. This property is basically due to the gel formation
potential and the ability to sorb organic molecules. Figure 5.30 illustrates
the breakdown and alignment of stacks of lamellae on remoulding and the
subsequent self-healing by rearrangement and re-establishment of interparticle
bonds. In a shear zone the same process takes place locally.

Figure 5.31 illustrates the thixotropic strength regain of very dilute smectite
grout used for sealing fractures. The shear strength was recorded by vis-
cometers and cone penetration tests. The self-healing leads to higher
strength for lower electrolyte contents because of the more homogeneous
microstructure implying more bonds per unit volume of the gels. Logically,
saturation with the most Ca-rich solution gave the lowest shear strength.
The strength regain goes on even beyond 100 hours.

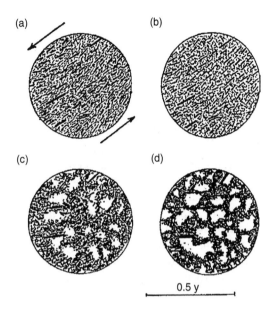

(a)

(b)

(c)

(d)

0.5 y

Figure 5.30 Transition from sol to gel form soft smectite after shearing (after M. Arnold).

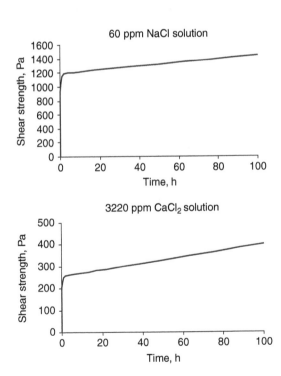

60 ppm NaCl solution

3220 ppm CaCl$_2$ solution

Figure 5.31 Thixotropic strength regain of smectite-rich clay with a density of 1110 kg/m^3 saturated with NaCl and CaCl$_2$ solutions, respectively.

5.7.6   Thermal behaviour

Influence of water content

The heat conductivity of clay depends primarily on the amount of water in the clay soil. Commercially available smectite-rich clay such as MX-80 is delivered with a water content (ratio of mass of water and solid matter) of 10–15%. This corresponds to 1–2 interlamellar hydrate layers in the clay soil. Compaction of the clay soil to a dry density of 1500 kg/m^3 gives a heat conductivity of about 0.75 W/m, K [18]. Various tests have shown that an increase in the degree of water saturation from 50% to 100% raises the heat conductivity by around 100%. Table 5.12 illustrates the influence of the degree of water saturation of MX-80 clay and of the Czech RMN clay as reported by the Czech Technical University in Prague, Centre of Experimental Geotechnics (CEG). The latter clay, intended as buffer material for a future HLW repository, consists of a mixture of 85 weight per cent of montmorillonite-rich Czech clay (Fe-dominated), 10% silica sand and 5% graphite.

A graphite additive has been proposed as a means to enhance the heat conductivity of buffer clays in several countries. Together with the heat-transfer capability of silica sand, the data in Table 5.12 demonstrates the effectiveness of this additive. Studies of mixtures of MX-80 with well-crystallized graphite show that the heat-enhancing ability of added graphite is due to the formation of a network of interconnected coatings of particle aggregates – much like the formation of quartz debris in Figure 4.21. These studies also showed that the degree of purity and crystallinity of the graphite is of considerable importance.

Influence of mineralogical composition

Comprehensive tests on a French candidate buffer clay FoCa7, which contains about 90% mixed-layer kaolinite/smectite, 5% kaolinite and 5% smectite in beidellite form, have produced results indicating a thermal conductivity

Table 5.12 Thermal data of RMN and MX-80 clays

Clay type	Dry density, kg/m^3	Degree of water saturation, Sr, %	Heat conductivity, λ W/m, K
RMN	1475	65	1.07
RMN	1597	99	1.22
MX-80	1570	70	0.90
MX-80	1700	98	1.14

of about 1.6 W/mK at 50–70% degree of water saturation for a dry density of around 1700 kg/m³. The microstructural constitution of the FoCa7 clay is similar to that of Ca-smectite.

Influence of stress and temperature

Applied effective stress and ambient temperatures are influential in the heat conductivity properties of unsaturated smectite clays. This is because a high effective stress results in a high density in the smectite clay with good inter-particle contacts and continuous films of adsorbed water. Under a thermal gradient, heat transfer will be aided by convection. This is most important at low degrees of fluid saturation.

5.8 Concluding remarks

The conceptual structural model of clay formed by compacting air-dry smectite grains with subsequent hydration implies that the clay does not become homogeneous, but contains interconnected voids filled with clay gels of lower density in comparison to the rest of the clay matrix. Taking micrograph-derived structural parameter data as a basis for application of flow theory for heterogeneous media, the calculated bulk hydraulic conductivity accords closely with the experimental values. This has been determined by the volume fraction and continuity of the permeable parts of the microstructure, that is, soft and medium-dense clay gels. The presence of soft parts explains why water under relatively high pressure can penetrate to a few centimetres depth in partly water-saturated clay, and why gas makes its way through channel-like paths in saturated clay.

The following major points should be noted:

- The derived microstructural models for compacted and hydrated smectite clay buffer provide data on the hydraulic and gas conductivities and the swelling pressure that agree well with experimentally determined bulk data – both for organic-free and organic-charged smectite clay over a large bulk density span. Microstructural parameters can be evaluated from micrographs for both clay types.
- The microstructural models described do not have sufficient resolution power to describe the most detailed processes involved in percolation, ion diffusion and shearing. However, since the structural variations on the gel scale are stochastic, statistically based models such as those described appear to be useful for developing theoretical models of greater detail.
- Smectites saturated with organic material such as $HDPy^+$ have a much more heterogeneous microstructural constitution than untreated clay.

The gel-forming potential is significantly lower. Organic cations can be adsorbed in the interlamellar space in a 'pillared' fashion, which allows interlamellar hydration. The stacks of lamellae are rigid and require much higher compaction pressure than 100 MPa to yield bulk densities exceeding about 1800 kg/m^3 after water saturation.

5.9   References

1. Yong, R.N. and Warkentin, B.P., 1975. *Soil Properties and Behaviour*. Elsevier Scientific Publishing Company, Amsterdam.
2. Anderson, D.M., Pusch, R. and Penner, E., 1978. *Physical and Thermal Properties of Frozen Ground*. Contribution to Chapter 2 of Geotechnical engineering for cold regions. McGraw-Hill, UK, 44–49.
3. Pusch, R., Muurinen, A., Lehikoinen, J., Bors, J. and Eriksen, T., 1999. Microstructural and chemical parameters of bentonite as determinants of waste isolation efficiency. European Commission Final Report Contract No. F14W-CT95-0012.
4. Pusch, R. 2002. The Buffer and Backfill Handbook. Swedish Nuclear Fuel and Waste Management AB (SKB), Technical Report TR-02–12. Stockholm.
5. Pusch, R. and Weston, R. 2003. 'Microstructural stability controls the hydraulic conductivity of smectitic buffer clay', *Journal of Applied Clay Sciennce*, Vol. 23: 35–41.
6. Pusch, R., Erlström, M. and Börgesson, L., 1987. Piping and erosion phenomena in soft clay gels. Swedish Nuclear Fuel and Waste Management AB (SKB). Technical Report TR 87–09, SKB Stockholm.
7. Neretnieks, I. and Moreno, L., 1993. 'Fluid flow and solute transport in a network of channels', *Journal of Contaminant Hydrology*, Vol. 14: 163–192.
8. Pusch, R., Moreno, L. and Neretnieks, I., 2001. 'Microstructural modelling of transport in smectite clay buffer', in *Proceedings of International Symposium on Suction, Swelling, Permeability and Structure of Clays*, K. Adachi and M. Fukue (eds), A A Balkema, Rotterdam/Brookfield.
9. Pusch, R., Karnland, O. and Hökmark, H., 1990. GMM – A general Microstructural Model for Qualitative and Quantitative Studies of Smectite Clays. Swedish Nuclear Fuel and Waste Management AB (SKB). Technical Report TR 90-43, SKB Stockholm.
10. Osipov, V.I., Sokolov, V.N. and Eremeev, V.V. 2004. *Clay Seals of Oil and Gas Deposits*, A A Balkema Publishers, Lisse/Atingdon/Exton (PA)/Tokyo.
11. Pusch, R., Ranhagen, L. and Nilsson, K. 1985. Gas Migration through MX-80 Bentonite. NAGRA Technical Report 85-36.
12. Horseman, S.T. and Harrington, J.F., 1997. Study of gas generation in MX-80 buffer bentonite. British Geological Survey, Edinburgh, WE197/7.
13. Kato, H., Muroi, M., Yamada, N., Ishida, H. and Sato, H. 1995. 'Estimation of the effective diffusivity in compacted bentonite', in *Proceedings of Scientific Basis for Nuclear Waste Management XVIII. Material Research Society, Symposium.*, *Vol. 353*, MRS, Pittsburgh.
14. Brandberg, F. and Skagius, K., 1991. Porosity, Sorption and Diffusivity Data Compiled for the SKB 91 Study. Swedish Nuclear Fuel and Waste Management AB (SKB) Technical Report TR 91-16, SKB, Stockholm.

15. Pusch, R., 1999. Mobility and survival of Sulphate-reducing Bacteria in Compacted and Fully Water Saturated Bentonite – Microstructural Aspects. Swedish Nuclear Fuel and Waste Management AB (SKB). Technical Report TR 99-30, SKB Stockholm.

16. Pusch, R. and Feltham, P., 1980. 'A stochastic model of the creep of soils', *Géotechnique*, Vol. 30, No. 4: 497–506.

17. Sutton, B.H.C. 1993. *Soil Mechanics. Longman Scientific & Technical*, Essex, UK.

18. Pusch, R. 2002. The Buffer and Backfill Handbook. Swedish Nuclear Fuel and Waste Management AB (SKB), SKB Technical Report TR-02-12. Stockholm.

6 Microstructural function of smectite clay in waste isolation

6.1 General

We have seen from the previous chapters that the role of clay microstructure in smectite engineered barrier systems is fundamental for effective isolation of waste. A major attribute of the clay material is its low permeability since this impacts directly on its transmittance properties – especially in respect to the transport of pollutants in the soil. Other attributes for these types of expansive clays include: (a) their potential for volume expansion (swelling) upon water uptake, (b) the ability to seal internal cracks, joints and openings in tightly confined situations because of the constraints against swelling and volume expansion, (c) development of a more homogeneous macrostructure as a result of the tightly constrained swelling process and (d) a high capability for partitioning of inorganic and organic contaminants.

Recent investigations of how clay-mixed waste can be isolated in abandoned mines [1] have shown how slow a clay liner becomes water-saturated and how this slow process affects the dissolution and migration of contaminants. In an underground repository, the ability of the surrounding rock (host rock) to provide water to the clay buffer that confines the waste canister determines the hydraulic boundary and thereby the conditions for the wetting of the clay. From the viewpoint of analysis and modelling of the performance of the buffer, this is a very critical part of problem conceptualization. Some of the necessary pieces of information required of the hydraulic boundary to determine or characterize the analytical and/or model loading boundary conditions include: (a) the manner in which water is delivered to the buffer (i.e. sporadic, continuous, intermittent, etc.) and (b) nature of, and accessibility to the water sources (point source, line source, finite, distribution of points etc). The low conductivity of argillaceous rocks, commonly used as the host rocks, suggest very strongly that complete water saturation of the clay buffers in the repositories used to contain highly radioactive waste will require many hundreds or even thousands of years.

The importance of the time for water saturation of initially non-saturated clay is particularly important for: (a) top liners of waste landfills where control and minimization of water entry into the wastepile are mandated requirements (b) for side and bottom liner and barrier systems since these are generally the last engineered barrier before release of the contaminants into the subsoil and (c) buffer and barrier systems in repositories designed to isolate high level nuclear wastes. The subject of saturation-time is a critical issue for repository situations, since evidence from prototype repository trials in different parts of the world show that complete saturation and permeation of the smectitic buffers could take well over hundreds of years. Before addressing this critical issue of saturation-time, we will provide a brief overview of the clay-engineered barrier system for landfills, to highlight the importance of the role of clays in development of barrier integrity. For a more detailed development of landfill liner and barrier systems, the reader is advised to consult the various textbooks dealing with waste landfills, and especially the various codes and regulations of different countries, since these regulations dictate the kinds of landfill liner systems to be used.

6.2   Clay for isolation of waste landfills

6.2.1   *General*

Waste landfill barrier-liner systems have one primary aim, that is, to contain and secure the wastes contained by the barriers. Design philosophies for landfill barrier-liner systems are guided, to a very large extent by the regulatory requirements of the country or region and economics. The 'garbage bag' principle adopted in construction of, and in operating, waste landfills satisfies the primary aim. It is generally acknowledged that since no impermeable protection system will remain forever impermeable, the accepted practice is to seek designs and methods that would minimize contamination of the surrounding ground by leachates, and especially the water phases (porewater, groundwater and aquifers) that constitute part of the ground. Figure 4.2 shows a simple implementation scheme for the garbage bag scheme, whilst Figure 6.1 shows a typical top and bottom-liner system for confining municipal solid waste (MSW) using clay soils as part of the top and bottom-engineered barrier system.

   As shown in Figure 6.1, one uses a multibarrier system for both the top- and bottom-liner systems. The bottom-liner system is designed to contain the leachates generated from the wastepile and the top liner is designed to control or even eliminate ingress of water to the waste mass. There are two schools of thought behind the control and/or management of water entry into the wastepile. One school of thought follows along the principle that by limiting the amount of water entering the wastepile, one limits the generation of leachates and therefore one will minimize potential leachate escape from the wastepile. The aim of this type of control is a dry wastepile garbage system. There are some very critical arguments concerning the ability of the liner systems to maintain their integrity over a protracted time

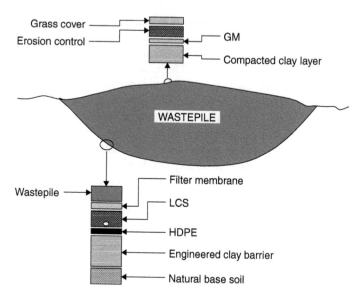

Figure 6.1 Typical general MSW landfill with top- and bottom-liner systems. HDPE refers to high density polyethylene membrane liner.

system – meaning that at some future time, leakage in the liner systems will inevitably occur.

Perhaps in recognition of the inevitability of leakages in any barrier system, and perhaps in the hope that accelerated processes can be used to some advantage, the other school of thought allows for water entry into the wastepile. This school argues that some water should be allowed to enter into the wastepile, and that leachate generated should be collected at the bottom and recycled into the wastepile to hasten chemical degradation of the waste materials. The amount of water entry into the wastepile will be dependent on the amount and quality of the leachate collected for recycling into the wastepile. To a certain extent, one tries to obtain a bio-chemical reactor system with this recycling technique.

The types of materials and designs for these top-liner systems vary considerably between: (a) different countries and regulatory requirements, (b) different kinds of wastes contained in the landfill and (c) different site conditions. In most countries, the regulations, criteria and standards (i.e. regulatory attitude) adopted by the regulatory body responsible for waste management and disposal (e.g. Ministry of Environment, Pollution Control Board, Environmental Management Bureau, Department of Environment, Environment Protection Agency, European Commission, etc.) will essentially dictate the type of landfill and barrier-liner system to be used. In the United States for example, MSW and hazardous wastes are regulated under the Resources Conservation Recovery Act (RCRA), including the Hazardous and Solid Waste Amendments (HSWA) to RCRA. Part 261 of Title 40 of the Code of Federal Regulations (40 CFR 261) provides the definitions for hazardous waste. Legislation applicable to MSW is contained in Subtitle D of RCRA and the regulations governing MSW landfills are covered in 40 CFR 258.

The two main types of regulatory attitudes are: (a) command, control and rectify (2CR) and (b) performance, monitor and rectify (PMR). The 2CR regulatory attitude basically specifies the type of liner systems to be used for landfills for containment of hazardous and non-hazardous wastes. For example, in the 2CR situation, single or double bottom liners with leachate collection systems (LCS) can be specified as required liner technology by 'regulations' for containment of non-hazardous and hazardous wastes, together with required groundwater monitoring accompanying installation of the landfill systems.

Figure 6.2 shows two general design schemes illustrating bottom-liner types for hazardous and non-hazardous waste containment. The LCS shown in Figure 6.2 consists of graded granular material in the layer, and a leachate collection pipe (white oval in the given figure) system. The lead detection system (LDS) shown in the right-hand section of Figure 6.2, consists of a layer of graded granular materials and a leak detection unit (gray oval).

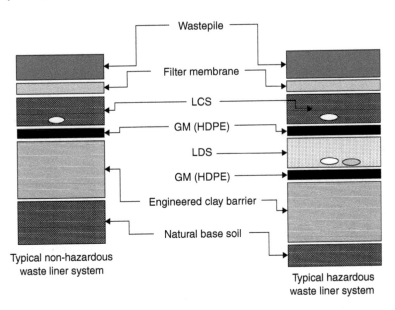

Figure 6.2 Typical landfill waste bottom-liner systems: (a) non-hazardous solid wastes on the left and (b) hazardous solid wastes on the right. Note that the white ovals in the LCS are collection pipes, and the gray oval in the LDS is the leak detection unit (adapted from Yong [2]).

It is not unusual for command and control types of regulatory specifications to seek performance requirements for LCS and LDSs. There is flexibility on the choice and use of technology for leachate collection. However, the system chosen must satisfy mandated requirements on the maximum leachate head permitted in the system (generally $\ll 0.5$ m). This essentially dictates the details of the leachate removal programme. Correspondingly, the LDS is governed by permeability constraints on the granular material in the layer (generally > 0.5 m) and a minimum permeability coefficient.

The PMR regulatory attitude does not mandate specific types of barrier-liner systems The PMR attitude requires the constructed facility to perform according to certain prescribed standards and to control leachate transport to conform to specified pollutant concentration trigger limits at specified distances and depths. By this means, control on performance capability of the landfill liner system is exercised. In essence, this type of attitude is a form of quality control on the constructed facility. If contaminants are detected in concentrations exceeding limit (trigger) values, at distances and time intervals previously specified as trigger values, this will constitute failure, and the facility must be rectified.

As in the case of the US RCRA legislation, the European Commission (EC) has recently defined the technical characteristics of waste landfills in the Council Directive 1999/31/EC. It specifies the following requirements for top, side and bottom liners:

- Landfill for hazardous waste: $K <$ E–9 m/s, thickness > 5 m.
- Landfill for non-hazardous waste: $K <$ E–9 m/s, thickness > 1 m.

Figure 6.3 shows some of the bottom-liner systems for some European countries. In all instances, it is the intention of the bottom clay liner to provide a secure and impermeable barrier against the transport of pollutants and contaminants. The various attenuation processes needed to make this clay layer function effective are well described in textbooks [3, 4] dedicated to such issues.

When it can be demonstrated that a bottom barrier system can adequately contain waste materials and their leachates without benefit of a membrane liner, the engineered clay layer that constitutes the liner must satisfy the PMR requirements all by itself. This situation is not uncommon for waste materials that are not considered to be hazardous or highly toxic. Assuming that a bottom clay liner is properly specified, one assumes that partitioning of the contaminants will serve to neutralize any leachate that finally makes its way out of the liner system into the subsoil system. Figure 6.4 shows an example of a mixed refuse dump of residues and ashes from Germany with clay top and bottom liners with a hydraulic conductivity $K <$ 5E–10 m/s. The dump must be located above the groundwater level.

Geomembranes (GMs), geosynthetic clay liners (GCLs), geonets (GNs), geotextiles (GTs) are some of the geosynthetics used in the liner-barrier systems that line landfills. Specifications for usage of the various types and

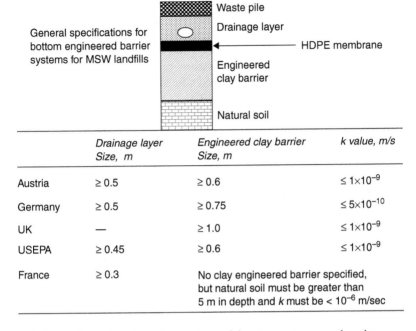

General specifications for bottom engineered barrier systems for MSW landfills

Waste pile

Drainage layer

HDPE membrane

Engineered clay barrier

Natural soil

	Drainage layer Size, m	Engineered clay barrier Size, m	k value, m/s
Austria	≥ 0.5	≥ 0.6	≤ 1×10⁻⁹
Germany	≥ 0.5	≥ 0.75	≤ 5×10⁻¹⁰
UK	—	≥ 1.0	≤ 1×10⁻⁹
USEPA	≥ 0.45	≥ 0.6	≤ 1×10⁻⁹
France	≥ 0.3	No clay engineered barrier specified, but natural soil must be greater than 5 m in depth and k must be < 10⁻⁶ m/sec	

Figure 6.3 General specifications for engineered barrier systems used as bottom-liner barriers to contain MSW (Adapted from Yong [3]).

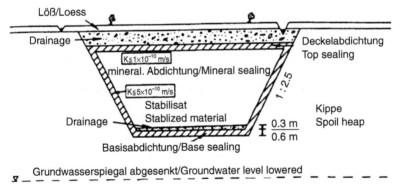

LöB/Loess

Drainage

Deckelabdichtung
Top sealing

K≤1×10⁻¹⁰ m/s

mineral. Abdichtung/Mineral sealing

1:2.5

K≤5×10⁻¹⁰ m/s

Stabilisat
Stablized material

Drainage

0.3 m
0.6 m

Kippe
Spoil heap

Basisabdichtung/Base sealing

Grundwasserspiegal abgesenkt/Groundwater level lowered

Figure 6.4 Schematic diagram of a refuse dump of buried type with incinerated ash, fly ash and sulphur components from gas cleaning. Seepage water is - collected at the deepest point and conveyed to a purification plant. The system requires monitoring and maintenance for the entire operational time.

for specific situations exist as design standards and as regulatory requirements. In the case of the engineered clay barriers shown in Figures 6.1 and 6.2, one finds that most often, specifications on clay material composition and properties refer to: (a) percentages of fines and coarse materials allowed,

(b) plasticity index range, (c) maximum permitted value for hydraulic conductivity and (d) minimum thickness of the clay layer. This contrasts significantly with the detailed material and performance specifications required for GMs used in the barrier-liner system. One has, for example, specifications or minimum values on thickness, tensile strength, durability, texture, gas and vapour transmission rate, solvent vapour transmission rate, etc. Typical types of GMs include: HDPE, very flexible polyethylene (VFPE) and polyvinyl chloride (PVC).

Saturation of engineered clay layers in top and bottom liners is a critical issue. Depending on the strategy or type of waste material contained, as discussed previously, except for the cases where bioreactor simulation is desired, it is claimed by one school of thought that it may not be advantageous to have a completely saturated clay top liner. The argument is made that once that occurs, it becomes difficult to protect against water entry into the wastepile unless an impermeable GM is used. Conservative design procedures anticipate that at some stage, imperfections, construction practice and degradation will all combine to produce eventual leaks into the wastepile. Hence, it is often prudent to design for a compact clay layer that offers the utmost protection against full permeation of water into the wastepile.

Similar concerns exist for the side and bottom-liner systems. We see from Figures 6.1 and 6.2 that the engineered clay liner is the last line of defence against release of leachates from the landfill. Leachates escaping from the landfill can only do so after penetrating the GMs and the engineered clay barrier or liner. Full saturation of the clay material would facilitate transport of contaminants to the underlying subsoil. An important question is therefore how quickly does the clay layer fully saturate, or more exactly, how does the wetting process take place? If it is sufficiently slow, it may mean that: (a) for the clay in the top-liner system, little or no water reaches the waste in several hundred years and (b) for the clay in the side and bottom-liner system, escape of fugitive contaminants to the subsoil will not be facilitated as long as the clay material remains partly saturated or dry.

A long-time factor for complete wetting of the liner material can be useful in meeting conformance requirements for buffer-liner performance. An example of this can be found in the Lithuanian concept for disposal of low-level radioactive waste. The natural attenuation of the radioactivity is set at 300 years and the top liner is designed to let no water through in this time period.

6.2.2   The wetting process of clay liners

The discussion to follow will focus on the performance and properties of the clay liners used for top, side and bottom-protection systems. It is understood that the function of these liners will not be materially affected whether they are overlain by HDPE and LCSs. This is because in either case we will need to confront the situation of leachate entry into this clay liner, no matter how long it takes to get there. As we have stated previously, the clay liner layer forms the last line of defence against transport of a pollutant load into the subsoil system.

We stress again that the design, and hence the construction, of top and bottom liners are subject to the various standards, criteria and requirements articulated by the regulatory bodies within the region responsible for the disposal and management of waste. Whether engineered top and bottom liners consist of zero membranes and LCSs, or single and/or double membrane liner systems as, for example, shown in Figure 6.2, is a matter that is decided by regulatory requirements. In the absence of such requirements, it is the responsibility of the facility owner and operator to provide a facility that would perform well, that is, without despoiling the environment and without being potential threats to public health and the environment. The discussions in the following sections concentrate on the properties, characteristics and performance of the last line of defence against transport of pollutants into the subsoil system.

The landfill case shown in Figure 6.4 is used as a focus for this discussion on the wetting process of clay liners. We will assume that for the top-liner system, the material is unsaturated and exposed to the elements. Water influx, by rainfall or other means, will be in the form of water inflow into major microstructural channels in the partly water-saturated clay. Water inflow will progress from the major channels to smaller channels, resulting in a downward-moving waterfront in the clay. However, the most important wetting process is diffusive migration of water with the suction (internal gradients) in the unsaturated part of the liner as the driving force.

The importance of pressure-induced wetting (i.e. via a hydraulic gradient) is stronger for liners with low clay contents than for clay-rich ones. We discussed this issue in Chapter 5, when we addressed the phenomenon of the critical gradient (Figure 5.5) using the double Mariotte tube experiment (Figure 5.4) to illustrate the significance of internal gradients in provoking water uptake in a swelling-type soil. The wetting rate can be very low even for liners with low-clay contents. Figure 6.5 gives the measured water content of samples taken at two different time periods from a compacted mixture of 10% Na bentonite (MX-80) and properly graded ballast material (10/90 mixture). The dry density and water content of the mixture were 1450 kg/m^3 and 2.5%, respectively. The soil was kept in a 0.3 m diameter and 2 m high column with a 0.1 m silt layer below the 0.9 m clay layer, over which 0.5 m sand and gravel had been applied. An artificial water level was maintained at 1 m height over the clay mixture. To avoid algae growth, a small amount of formaline was added to the water.

As shown in Figure 6.5, the rate of wetting is so slow that complete water saturation would not be expected until after many decades. The curved shape indicates that diffusion is the major wetting mechanism, and that the 'bump' of the central parts of the curves is due to pressure-induced inflow of water. The results portrayed are from experiments started in 1987 with sampling made in 1991 and 1993.

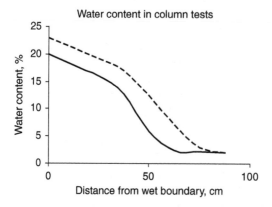

Figure 6.5 Water content of 0.9 m column of 10% Na bentonite (MX-80) and 90% ballast in a mock-up test starting in 1987, with samplings made in 1991 (lower curve) and in 1993 (upper curve).

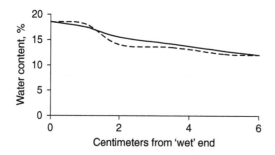

Figure 6.6 Comparison of predicted and measured water content of FIM clay with 1900 kg/m³ dry density. The smooth curve represents prediction with D = 3E–10 m²/s whilst the irregular curve represents the results from the experiments.

The rate of diffusive wetting, which can be evaluated from such tests by curve fitting procedures, will yield diffusion coefficients generally from E–11 to E–9 m²/s depending on the smectite content and density. For FIM (Friedland) clay, used here as a reference clay of intermediate quality and relatively low cost, the wetting is typically diffusive with the diffusion coefficient being about 3E–10 m²/s for a dry density of 1700 kg/m³ (Figure 6.6). Such a high density can only be obtained with difficulty for pure clay materials. For the easily achieved dry density of 1200 kg/m³ for FIM clay, the diffusion coefficient is about E–9 m²/s.

The risk of intrusion of heterogeneity as a factor, especially in mixtures prepared with low clay and water contents for laboratory testing makes it necessary to use large samples. This will provide REVs that would allow for a more rational microstructural characterization. Figure 6.7 shows a medium-sized

Figure 6.7 Oedometer with 250 mm diameter and removable bolts for sampling during the test.

oedometer specifically designed to allow for proper testing of the influence of the microstructural features of the clay soil on its hydraulic performance.

Since the hydraulic gradient or water pressure on liners will affect the rate of wetting, it is necessary to undertake wetting/percolation tests under different piezometric pressures. The aforementioned special shape of the experimental wetting curve caused by water pressure superimposed on the diffusion-wetting is illustrated also in Figure 6.8 for the FIM clay with a dry density of 1200 kg/m³. The agreement between the predicted and measured wetting rates, assuming diffusion-wetting to be characterized by $D = $ E–9 m²/s, is obvious – with the exception of the pressure-induced 'bump'. Using this diffusion coefficient, theoretical calculations show that a liner of FIM clay with 0.5 m thickness will require 100 years to become water-saturated. If we increase the thickness to 0.8 m, the time period will increase to 1000 years. Everything being equal, we would not expect the wastepile to generate sufficient leachates in this time period to threaten the bottom engineered barrier and clay liner system. This clearly demonstrates the advantage of making top liners very tight [5].

6.2.3 Percolation (permeation) of liners

General

The rate of percolation of the top clay liner, which may require tens to hundreds of years to become water-saturated, determines the downward transport of ions released from the underlying waste to the bottom-barrier system, and finally through the bottom clay liner. All else being equal, the

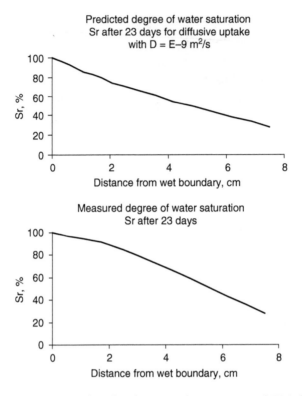

Figure 6.8 Predicted and measured wetting rate of FIM clay with 1200 kg/m³ dry density exposed to a constant water pressure of 80 kPa.

percolation rate is controlled by the composition and density of the upper liner. A well compacted high density clay barrier is most desirable. The compatibility of the bottom clay liner with the chemistry of the leachate is a critical element of liner durability. The rate of percolation of waste piles determines the degree of contamination of drinking water in downstream wells, and prediction of this rate is critically important for proper location of waste landfills and for constructing the low-permeable barriers [5]. Diffusive migration through such barriers of ions released in the course of the percolation is also an important transport mechanism.

Impact of density and porewater chemistry

The higher the density of clay layers, the better their capability to serve as waste isolation barriers. However, the high swelling pressure that accompanies it can cause problems if the swelling pressure exceeds the effective overburden pressure. If this occurs, the volume expansion will cause upheaval, softening and an increase in hydraulic conductivity of the liner. Although local conditions and requirements for the landfill may not allow the landfill to be sited such that the top-liner system will be below the frost line, efforts need to be made to ensure protection against the forces of weathering. This is particularly critical for the top-liner system since this layer will be subject to erosive forces, wet–dry cycles and also freeze–thaw cycles. For optimum performance of the side and bottom-liner systems, one needs to select a suitable density, composition and thickness of clay liner that would: (a) be chemically compatible with the chemistry of the leachate entering the clay layer, (b) provide the desired contaminant partitioning capability, (c) provide the design permeability constraint and (d) be durable.

Soil composition, soil microstructure and soil-water interaction are integral components in the control of soil permeability, especially when the soil is exposed to various kinds of contaminants in the leachate stream penetrating the clay layer. For a non-swelling soil such as kaolinite, the evidence shows that it is relatively insensitive to moderate variations in leachate chemical composition. The ranges of permeability values obtained from rigid and flexible wall permeameter tests of kaolinite clay with water and also with a Pb-leachate, show values of K between 1.3E–9 and 6.5E–9 m/sec [6]. Soils composed predominantly of illite show moderate decreases in permeability with decreasing salt concentration. The permeability of clay soils with some significant proportion of montmorillonite (e.g. MX-80) will be susceptible to the effects of changes pf salt concentrations in the porewater. In addition, the species of exchangeable cations can also impact directly on the microstructure of the soil structure by altering interlamellar and interparticle forces. Interactions between the water molecules and the cations on the clay surface and in the diffuse double layer will influence the permeability characteristics of the soil. Permeability will be decreased by the restriction on the mobility of the water forming the hydrogen shell around the cations. Another possible factor to be considered in fluid flow is the degree to which the cations alter the actual behaviour of the hydrates in interlamellar space. The extent of the influence from these factors on permeability testing (i.e. fluid flux), depends on the size, charge and concentration of ions in the pore solution.

The dissociation of the hydroxyl (OH^-) groups exposed on the clay crystal surfaces and edges is sensitive to the pH of the immediate environment. The higher the pH value, the greater is the tendency of H^+ ions to go into solution, and hence, the greater is the net negative charge of the clay particle. In addition, aluminium exposed at the edges of the clay crystals, is amphoteric and ionizes positively at low pH values, and negatively at high pH values. A low pH promotes positive edge to negative surface interactions of clay particles. This will result in a flocculated structure, and hence, an increase in permeability. A high pH promotes a negative charge on the edges and favours the development of a more dispersed fabric and a resultant decrease in permeability.

Leachates containing inorganic acids may dissolve some constituents (e.g. aluminium, iron and alkali metals) that comprise the clay soil structure. The clay soils used for liner systems have clay minerals that contain aluminium in large quantities that are highly susceptible to partial dissolution by acids. The solubility of clays in acids varies with: (a) the nature and concentration of the acid, (b) the acid to clay ratio, (c) the temperature and (d) nature of the anion in the acid. When the anion in the acid is about the same size and geometry as a clay component, even weak acids can dissolve clays under some conditions. Results from studies [7] at high (boiling point) temperatures show that the percent of solubilization of aluminium was 3% for kaolinite, 11% for illite and greater than 33% for montmorillonite. Permeation with acids leads to a one order of magnitude increase in the hydraulic conductivity of the montmorillonite specimen in comparison to that determined using water as a permeant. The differences or modifications in conductivity in relation to chemistry of the permeating fluid can be attributed to:

- extraction of lattice aluminium ions from the octahedral sheets of the clay minerals;
- ion exchange on the surface of the clay minerals due to replacement of naturally adsorbed cations of lower valence (Ca^{2+}, Mg^{2+}, Na^{2+}, K^+) by the extracted aluminium ions which has a valence of 3 and hence, a reduction in the thickness of the diffuse double layer;
- an increase in effective pore space and a decrease in the tortuosity factor, resulting thereby in a higher permeability coefficient.

In respect to leachates containing inorganic bases, mineral dissolution is also a very significant problem. Inorganic bases have the ability to increase the net negative charge on the clay surfaces, and the capability to dissolve silica. Smectitic clays have minerals that contain large quantities of silica in their tetrahedral sheets that are susceptible to particle dissolution by inorganic bases. As has been stated repeatedly, replacement of exchangeable sodium by calcium ions will result in a significant increase in the permeability of these types of clays. This is due not only to the reduction in the double diffuse-layer (DDL) interactions, but also the formation of quasi-crystals between the exchangeable calcium ions and a pair of opposing siloxane ditrigonal cavities [8]. This makes it important to know exactly what type of swelling clay one needs to specify for use as a clay liner material.

The replacement of monovalent sodium ions by trivalent ferric ions has a significant effect on the permeability coefficient of bentonite – with an increase of more than 30 times. The increase in permeability is due to the combined effects of: (a) reduction in the diffuse double layer thickness due to replacement to the monovalent sodium ions by trivalent ferric ions and (b) hydrolysis of the trivalent ferric ions to form iron hydroxy species that become coatings around the quasi-crystals – affecting their aggregation (and hence the microstructure) by electrostatic bonding.

Organic chemicals escaping from a landfill will create some very interesting issues when these chemicals come into contact with the clay material in the barrier-liner system. Permeation of the clay layer by an organic chemical will cause changes in interlayer spacing. For organic chemicals with dielectric constants lower than water, the individual clay particles (composed of stacked lamallae) will contract as a result of a thinner interlamellar spacing, thus providing an opportunity for the clay particles to orient themselves. This could result in changes to the permeability of the clay. The effects can be very significant for swelling soils such as bentonites, but of lesser significance for non-swelling soils that are compacted to a high density.

Movement of organic molecules in the clay layer will be by diffusion and advection through the macropores. Along the way, partitioning occurs between the aqueous phase and soil aggregates. Molecules weakly adsorbed by soil aggregates will move more quickly through the aqueous channels. Hydrophobic substances such as heptane, xylene and aniline are highly partitioned, and would consequently be expected to develop resultant soil-heptane, soil-xylene and soil-aniline permeabilities lower than those of soil-water and soil-acetone.

There is a relationship between the hydraulic conductivity K and the octanol/water partition coefficient k_{ow}. This octanol/water partition coefficient accounts for the tendency of permeant molecules to escape from the aqueous phase. By and large, the permeability of a clay soil in respect to a liquid permeant decreases as the log of its octanol/water partition coefficient increases. Because the octanol/water ratio is a measure of escaping tendency of the organic material from water, those substances least compatible with water should move most slowly through the soil. The more positive the value of the octanol/water ratio, the lower is the value of permeability. In other words, the more hydrophilic the organic chemical, the more rapidly it moves through the clay soil.

Chemical analyses of pore fluids of soils permeated with organic chemicals have shown notable decreases in cation concentration in the porewater. The decrease in cation concentration could lead to an increase in the repulsive forces between particles, the outcome of which is soil particle dispersion and a decrease in permeability. In general, the repulsive energy between soil particles increases with an increase in the octanol/water partition coefficient, resulting thereby in a corresponding decrease in the permeability of the soil.

Experimental tests show that there is a good correlation between the permeability of a clay soil permeated with an organic chemical, and its dielectric constant. The greater the dielectric constant, the greater is the permeability. This correlation should be expected since the dielectric constant is an approximate measure of a liquid's hydrophobic or hydrophilic character. Substances with high dielectric constants tend to be hydrophilic and can therefore be expected to move more quickly through the aqueous channels in the clay. Low dielectric substances will be adsorbed and in consequence will demonstrate retardation in their movement through the soil.

As a general rule, as the molecular weight of the organic chemical increases, the clay soil permeability, in respect to the organic chemicals, decreases. This is because more water molecules will be displaced, and also because the large molecules have more points of contact with the active clay surfaces. In the adsorption of long chain molecules, van der Waals interactions are important. This is due to the fact that these forces are additive and tend to reorient the organic molecules for maximum contact points with clay surface. This explanation is supported by the observation that molecular weight can be used as a measure of organic substances' hydrophobicity. The greater the molecular weight, the higher is the tendency of organic substance to be hydrophobic. Movement of such molecules can therefore be expected to be slower through the aqueous channels in a clay soil.

Percolation (permeation) of the upper clay liner

Following water saturation of the upper clay liner, percolation occurs under the hydraulic gradient produced when a pressure head develops in the overlying drain layer. Prediction of the percolation rate is a simple exercise. However, the specification of pressure heads and their persistence requires estimation of the water balance. This will be based on statistical precipitation data and prediction of future impact by man and by climatic changes. As a very conservative case, one can assume that the water level in the drain layer on top of the upper clay liner is maintained at a height over the clay liner that corresponds to 100% of the annual precipitation. This is assumed to be 1 m in the following discussion.

Using the Darcy model as a basis for calculations, the percolation rate is determined as follows:

$$v = K \times i \tag{6.1}$$

where v is the flow rate (m/s), K is the hydraulic conductivity (m/s) and i is the hydraulic gradient (m/m).

Taking K as E–12 m/s, which is representative of a high-quality Na smectite clay with a dry density of about 1100 kg/m^3, and the thickness of the clay layer as 1 m, one has $i = 1$ and $v = $ E–12 m/s [9]. This means that a permeating water molecule moves about 1 mm in 300 days. At this rate, it would take 1000 years for it to migrate through the 1 m thick clay layer. For a less tight clay such as the FIM clay with about 45% swelling clay minerals, the calculations will show that the conductivity after water saturation is still less than E–10 m/s [9]. For a dry density of 1500 kg/m^3 for the FIM clay, which is a density that is easily obtained in the field, the time for through-flow of water is calculated to be about 10 years. If we take into consideration intermittent precipitation and drought, where the average piezometric height in the drain layer covering the liner would not be more than 1/3 m, we will calculate a through-flow in about 30 years after water saturation. The lesson we learn from this is that percolation of the upper liner is very strongly retarded when we use a smectite-rich clay material. In addition, considerable retardation can be obtained by using a high drainage capacity of the overlying material such as a coarse-grained layer, since this will alleviate the water pressure. The average amount of water flowing through the upper liner is the product of time, hydraulic gradient and hydraulic conductivity. This gives us about 1 litre per year per square metre for the FIM clay case after complete water saturation.

Microstructural aspects on the use of liners made of clay/ballast mixtures

The very low hydraulic conductivity of liners that is usually desired requires that the void size is minimized by very effective compaction, or that the content of smectite clay is high. In practice, cost is a determinant. This is unfortunate because experience shows that this means using low clay contents and inefficient compaction. Accordingly, in the face of reality, we need to seriously consider the question of distribution and density of the clay component in mixtures discussed in Chapter 5.

The following applies:

1   For the common bulk densities and grain size distributions of mixtures of smectite clay and ballast, a clay content of about 10 weight per cent is required in practice for effective filling of the voids between ballast grains. This takes into account that perfectly uniform distribution of the soil components cannot be obtained under field conditions.

2   For common ballast grain size distributions and compaction efforts, a clay content of 10–15% means that the clay component, which controls the bulk hydraulic conductivity, will not be higher than about 1600 kg/m^3 irrespective of the bulk density. This gives a bulk conductivity of no less than E–10 m/s even for very effectively compacted top liners saturated and percolated by low-electrolyte water. For bottom liners, porewater chemistry determines the bulk conductivity. This can be orders of magnitude higher.

3   For the same compaction energy, an increase in clay content beyond 15–20% reduces the density of the clay component, resulting in a net reduction in bulk density of the mixture as illustrated by Figure 6.9.

4   Clay contents lower than 15–20% will make the mixtures somewhat heterogeneous with respect to distribution of clay and therefore sensitive to piping and erosion. Water will find the paths of least resistance, which are represented by the narrow space between adjacent ballast grains and by the least dense parts of the expanded clay component (Figure 4.17).

5   The heterogeneity and low density of the clay component for clay contents lower than 10–20% makes it very sensitive to an increase in

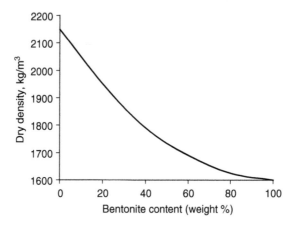

Figure 6.9 Relation between smectite (MX-80) content and dry bulk density for 100% Proctor compaction using common Fuller-graded ballast as the non-clay component.

electrolyte content, especially when Ca is the dominant cation. A minimum clay content of 30% is believed to eliminate the problem.

6   Top liners may undergo considerable strain if the waste mass is compressible and undergoes differential settlement.

All these statements are well known to geotechnical engineers and have led to a comprehensive development of theories and practical rules. Prominent amongst these are practical methods to: (a) determine how grain size distribution should be selected (b) implement mixing and compaction of clayey materials for lowest possible conductivity (c) obtain highest possible erosion and piping resistance and (d) obtain best self-sealing potential after mechanical disturbance. Microstructural modelling is a key issue. It has been shown that the so-called D_5 parameter, which is the percentage of particles representing the fifth percentile of the grain size curve, is a determinant of the hydraulic conductivity. The background is as follows.

Starting from Darcy's basic flow equation, one can express the unit flux q in terms of mass of water flowing perpendicularly through a given area of an isotropic porous medium as a function of seven characteristic parameters [10]:

$$q = f(i, \gamma, \mu, D_x, S, \eta, n) \tag{6.2}$$

where i is the hydraulic gradient, γ is the density of fluid, μ is the viscosity of fluid, D_x is the controlling grain size, S is the gradation factor expressing the shape of the grain distribution curve, η is the grain shape factor and n is the porosity.

From dimensional analysis, the following expression can be derived:

$$q = i(\gamma/\mu)k_0 \tag{6.3}$$

and

$$K = (\gamma/\mu)k_0 \tag{6.4}$$

where K is the experimentally determined hydraulic conductivity, k_0 is the 'specific permeability' of dimension (length)^2 describing the geometry and size of the void network.

k_0 can be expressed as:

$$k_0 = \beta_\alpha \cdot D_a^2 \tag{6.5}$$

where β_α is the dimensionless factor describing the geometry (but not the size) of the void network and D_α^2 is the measure of the cross-sectional area of the average pore channel.

Figure 6.10 shows the relationship of log k_0 and log K versus log D_5. The figure shows that the relationships form bands for $C_u = 1–3$ and $C_u > 3n$ ($C_u = D_{60}/D_{10}$), the first one representing relative uniform grain size and the other, flatter grain size curves in ordinary grain size diagrams. The implication of the theory is that C_u is not a determinant of the hydraulic conductivity but that D_5 is a key parameter, implying that the few weight per cent of the finest material most significantly affects the hydraulic conductivity. By choosing different values for α in D_α, the bands in the figure will change their relative positions. Various studies have shown that one can take $\alpha = 5$, and that D_5 is the conductivity-controlling grain size largely independently of the size distribution of the rest of the mineral assemblage – provided that the gradation curve is not discontinuous. Laboratory and field experiments with mixtures with very low D_5 validate the theory (cf. Figure 6.11).

To use the graphical information in Figure 6.10, the following procedure is required:

1   Determine the gradation coefficient C_u. If $C_u > 3$, use the right side of the figure.
2   Determine D_5 from the grain-size distribution curve.
3   Identify D_5 on the horizontal axis and read the hydraulic conductivity on the right vertical axis.
4   The following example can be used to check on the procedure: for $C_u = 12$ and $D_5 = 0.1$ mm; we will obtain $K = 0.006$ cm/s $= 6$ E–5 m/s.

Taking for instance $D_5 = 0.1$ mm, the average conductivity from Figure 6.10 is about E–4 m/s, while for $D_5 \sim 0.3$ mm it is E–3 m/s.

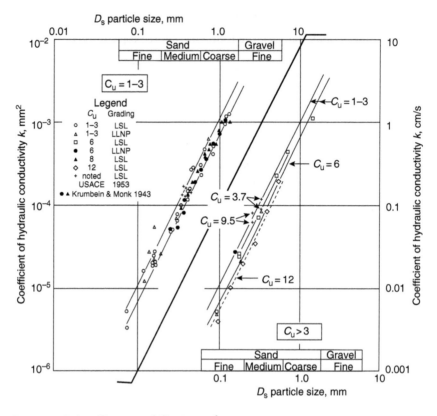

Figure 6.10 Specific permeability in mm^2 versus D_5 grain size in mm [10].

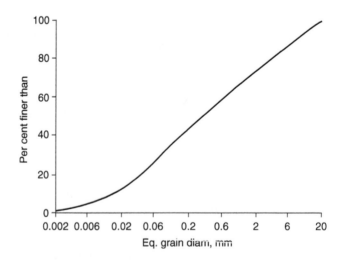

Figure 6.11 Grain size distribution of mixture of moraine and 4% very fine Na bentonite yielding the very low hydraulic conductivity of E–11 m/s. $D_5 = 0.006$ mm.

Extrapolation suggests that the conductivity may be as low as E–5 m/s for $D_5 \sim 0.05$ mm. The conductivity calculated by use of the diagram in Figure 6.9 fits fairly well with Swedish and Finnish investigations of ballast materials [9].

The significance of the role of the finest material in permeation is also demonstrated when we apply Poiseuille's law Eq. (6.6). Soils with very small particles also have small r-values, yielding low flow rates and fluxes as follows:

$$v = \pi p \cdot r^4/8l\eta \qquad\qquad (6.6)$$

where v is the flow rate, p is the pressure, r is the channel (void) diameter, l is the channel length and η is the viscosity of fluid.

*Microstructural issues in the use of liners made of
clay/ballast mixtures*

The difficulties in achieving acceptable performance of liners of mixtures of smectite and ballast have led to a greater consideration of the use of natural clays with a smectite content of at least 50%. Taking the FIM clay as an example, the following advantages are noted:

- Homogeneity is much better in comparison with mixtures because the clay component is represented by grains of all sizes.
- Expandability, and hence self-sealing ability, is better because the clay component is represented by grains of all sizes.
- Hydraulic conductivity is lower than for mixtures of 30% smectite-rich clay (MX-80) and well-graded ballast for one and the same density.
- Cost may be lower than for mixtures of 30% smectite-rich clay (MX-80) and well-graded ballast, since mixing operations and interim storage of the different components is eliminated.

The physical properties of clay/ballast mixtures and fillings of dried and ground natural clay materials need proper consideration. This requires a better understanding of the function of particles of different size and hydration potential in laboratory and field compaction. Traditionally, high densities of compacted soil fillings mean preparation of the materials by adjusting water content to 'optimal'. However, this rule may not be followed if the smectite minerals content is 10–20%. For such materials, experience shows that the soil components should be very dry if effective compaction can be obtained to force the particles together. This is illustrated by the results of laboratory compaction tests shown in Figure 6.12. For a low water content (2–3%), compaction by heavy punching produces the highest maximum density (MCV). The 100% saturation ($S_r = 100\%$) represents the highest density that can theoretically be achieved.

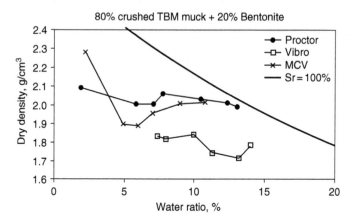

Figure 6.12 Compaction curves of a mixture of 20% MX-80 clay and crushed rock using different compaction techniques. The conventional 'optimum' water content is 8–10%. However, the highest densities were obtained for mixtures with very low water contents.

The explanation for the two density maxima phenomenon is obvious when we take into account the microstructural conditions. When water is added to a mixture of ballast and smectite clay, the clay hydrates and forms coatings of the ballast grains. As a result of this, the coated grains will not move easily, that is, not until the amount of water is high enough to make the clay a lubricant. At this stage, one arrives at the conventional 'optimum' water content. For higher water contents, build up of porewater overpressure will prevent effective approach of the ballast grains. At very low water contents, clay granules behave like ballast grains and will combine with the ballast grains to move into closest possible layering. This will give us the most homogeneous microstructural constitution of the clay component after wetting of the compacted mixture. Consequently, we will obtain a lower hydraulic conductivity in comparison to compaction at higher water contents. Practical application of the technique, which is termed 'dry mixing and compaction', has yielded dry densities of up to 2300 kg/m³ at field compaction of mixtures of 10% MX-80 clay and well-graded ballast material. In terms of microstructure, this corresponds to a dry density of the clay component of more than 1100 kg/m³.

6.2.4 Mechanical behaviour of liners

Differential settlement

A major advantage of using clay-rich material for constructing liners is that they can accept mechanical strain much better than mixtures of

ballast material and clay with clay contents lower than 20–30%. This is demonstrated by laboratory studies and ongoing field tests of mixtures with 12% FIM clay and graded ballast in the form of ash (Figure 6.13). Laboratory tests of FIM clay mixtures with dry densities of 1000–1430 kg/m³ will be taken as examples of the impact on shear strain on the hydraulic conductivity of clayey soils.

In the laboratory study conducted using shear boxes schematically shown in Figure 6.14, sample shearing was undertaken at a constant strain rate of 0.5 rad in 1 hour until an angular strain of 30°was obtained. The hydraulic conductivity was determined before and after shearing. From FEM analyses, we can show that a plasticity condition (i.e. plastization) occurs at the contact between the cell halves, at the rim, for a few millimetres of relative displacement [11]. This means that the shear stress τ in the central intact part of the sheared section can be described as follows:

$$\tau = [F - \tau_f(A - A_s)]A_s \qquad (6.7)$$

where τ_f is the shear strength represented by the plasticized part, A_s is the cross section area of the central intact part.

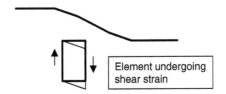

Figure 6.13 Shear strain by differential settlement of a top liner. The overburden consisting of vegetation, erosion-protecting material, drain layers and silt filters is not shown.

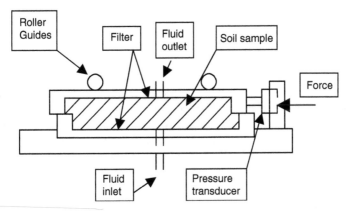

Figure 6.14 Direct shear apparatus used for the laboratory tests. The cylindrical cell has an inner diameter of 75 mm and the sample is 20 mm high.

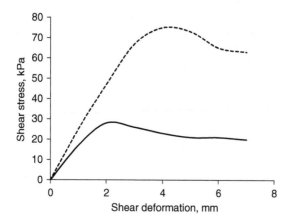

Figure 6.15 Evaluated shear stress/strain behaviour yielding a maximum shear stress of 28 kPa for the softest mixture (lower curve) and 75 kPa for the densest.

The shear strain γ of the soil element before development of complete shear failure is:

$$\gamma = 3\Delta/D_s \qquad (6.8)$$

where Δ is the relative displacement of the cell halves, D_s is the diameter of the central intact part of the sample.

The larger Δ is, the smaller is D_s and γ therefore becomes large when plasticization becomes significant (cf. Figure 6.15). For 70% plastization D_s is $0.3D$ and γ slightly more than 01Δ, corresponding to about 30°. The microstructural impact brought about by this large strain is significant. This can be seen as an alignment of a significant proportion of clay aggregates in the direction of the shear plane. The uniformity in distribution of clay aggregates in this zone produces a drop in the hydraulic conductivity with a no-volume change condition.

The failure envelope for the direct shear test is shown in Figure 6.16. The results show that there is no cohesion intercept, meaning that the shear strength is zero if the upper surface is unloaded, and that the friction angle is about 17°.

Evaluation of the hydraulic conductivity before and after shearing gave the data in Table 6.1. They confirm the expected drop in hydraulic conductivity after shearing of the non-dilatant mixture of smectite-poor clay and ash, which had no chemical impact on the clay component in the short testing time.

Slope stability – short term

Top liners have slopes that need to be stable for the entire operational life of the landfill or waste containment system. This can be more than

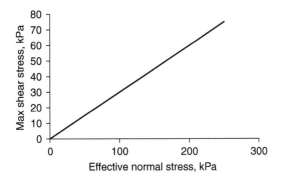

Figure 6.16 Failure envelope in Mohr–Coulomb representation of mixture of 12% FIM clay and ash. The friction angle is 17° and there is no cohesion intercept.

Table 6.1 Hydraulic conductivity *K* of mixtures of 12% FIM clay and ash before and after shearing

Density of water saturated sample, kg/m³	K, m/s before shearing	K, m/s after shearing
1650	8E–9	4E–9
1900	3.2E–10	2.8E–10

300 years, depending on the local regulations and legislation. The stability of the slopes depends on the shear strength and on the capability of the material to sustain strain. These matters are part of the geotechnical considerations in the design of liners, and the literature contains numerous examples of successful, and some suspect cases of technical solutions based on the use of smectite clay. We will examine a basic case and discuss the most important issues for such slopes.

The tight part of a top liner is commonly not more than one metre thick, and slope failure in the form of sliding along a plane slip surface will take place if the Mohr–Coulomb cohesive strength (intercept *c*) and the friction angle ϕ are inadequate. For the basic case shown in Figure 6.17, knowledge of the strength parameters together with the usual soil mechanics analytical procedures will give a factor of safety *F* against failure as:

$$F = \tan \phi / \tan \beta \qquad (6.9)$$

where β is the slope angle.

Following along the lines of traditional soil mechanics and foundation engineering, the safety factor, that is, the ratio of the shear strength and the

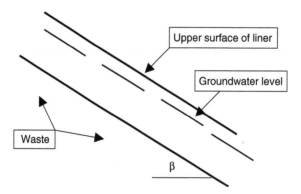

Figure 6.17 Section of sloping top liner with 'upper' groundwater level above the tight clay-based layer.

shear stress, along an assumed kinematically possible slip surface, should be at least 1.5. This criterion can be readily fulfilled for excavated slopes in frictional materials and for many illitic clays. However, smectitic clays will not be able to conform to this criterion since the friction angle for pure Na and Ca smectites is only about 10–15°. With such a restrictive friction angle for smectitic clays, the stable slope angle would have to be less than about 6–10°.

There are designs and constructed slopes in practice with top-liner slope angles of about 20°. If smectite clays are to be used, these would need to be smectitic clays with high friction angles. Candidate materials include FIM clay, or mixtures of clay and ballast, or ash deposits obtained by mixing clay and waste – as presently being tried in Sweden. Taking the earlier example of 12% FIM clay mixed with 88% ash with a friction angle of 17°, the safety factor 1.5 would still limit the slope angle to about 11°. This is not acceptable for most cases. Geotextiles have often been proposed as an aid in stabilizing steeper slopes. This should be considered as a short-term solution since the risk for chemical degradation must be confronted and overcome if long term application is to be sought with such materials. In practice, one needs to consider acceptance of lower factors of safety, for example, $F = 1.1–1.2$. For the case with FIM clay mixed with ash $F = 1.15$ would correspond to a slope angle of 15°.

Slope stability – long term

Time dependent shear strain (creep) is a significant factor that must be considered by designers of waste landfills. The mechanisms leading to creep and their microstructural implications have been briefly discussed in Chapter 5. Experience shows that for shear stresses in the interval 1/3–2/3 of the conventionally determined shear strength, the creep rate will attenuate with time according to some exponential law. From the viewpoint of thermodynamics and stochastic mechanics, the creep attenuation–time relationship would be described by a *log time* function. For lower shear stresses, the attenuation of strain rate is quicker. Referring to the creep theory in Chapter 5, the formation of microstructural slip units is insignificant for low stresses, whereas for stresses of high intensity, a critical number of microstructural slip units will be produced, leading thereby to an accelerating creep rate followed by complete failure [12].

Application of intermediate stresses will lead to the formation of local regions of slip. Continued slip will result in local stress relaxation and an increase in the heights of the energy barriers for subsequent activated jumps. This is akin to work-hardening in metals. The microstructural meaning of the creep equations for intermediate stresses is that each transition of a slip unit, between consecutive barriers, adds a certain small contribution to the bulk strain, the integration of which over the energy spectrum $N(u,t)$ gives the expression for the shear strain rate (cf. Chapter 5):

$d\gamma/dt$ proportional to the integral between the upper
  barrier level u_2 and the lower level u_1 of $N(u,t)du$
yielding: (6.10)
$d\gamma/dt$ proportional to $(1 - t/t_0)$
  with $t < t_0$ as boundary condition.

For stresses lower than about 1/3 of the conventionally determined shear strength, one expects that slip units are effectively hindered by the higher energy barriers confronted. Self-repair occurs in the stress relaxation process. Creep slows down quickly. The theoretical strain as a function of time is given as:

$$\gamma = \alpha - \beta t^2 \text{ (with the condition } t < \alpha/2\beta) \qquad (6.11)$$

that is, the creep starts off linearly and then dies out.

The conventional third creep phase is reached when the stresses are higher than about 2/3 of the conventionally determined shear strength. At this stage, the initially decreasing creep turns into a strain that is linearly proportional to time. The strain rate is too high to allow for self-repair of the damaged microstructure. Strain becomes faster and faster, and ultimately failure occurs. The appearance of the creep curve in this phase is of the type shown in Figure 6.18.

Safe design of top liners requires one to perform two types of stability calculations, (1) short-term conditions and (2) long-term conditions considering creep. For the short-term case, one has to evaluate the stability factor, that is, the ratio of the shear strength and the shear stress, for the respective component. This refers to the uppermost soil with vegetation,

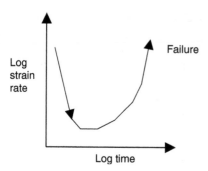

Figure 6.18 Generalized shape of the creep rate curve in double logarithmic diagram at a critically high shear stress.

drainage layer, silt filters and clay-based layer. Long-term conditions are seldom considered, and stability problems in the future can be expected.

Liquefaction

Liquefaction is a stability issue that needs to be considered. Liquefaction may be a serious problem for any water-saturated low density soil fill. Seismically induced shearing will cause contraction and development of a *porewater overpressure* (in conventional soil mechanics, this is often referred to as *excess porewater pressure*) that can reduce the effective pressure to a critically low level, particularly in slopes with significant inclination. For clay-based soils, densities at water saturation exceeding about 1800 kg/m^3 cause no problem [13].

6.2.5   Overall isolation performance of the system of liners and waste

Relative importance of top and bottom liners

For the situation of an initially dry wastepile in the landfill, hydration of the bottom clay liner does not commence until water has moved through the waste mass and penetrated the synthetic liner overlying the clay layer. This assumes that the bottom barrier design includes an LCS and a synthetic liner overlying the engineered clay layer, as shown in Figures 6.1 and 6.2. If only an engineered clay barrier is used (without benefit of synthetic liner and leachate collection), application of the previous calculation procedures for penetration of water through the top liner will show that water will not reach the bottom clay layer until after several tens of years or even centuries following completion of the landfill.

As we have stressed in the previous sections and chapters, density is a key parameter for the transmissivities of the clay liners at top and bottom, and that it (i.e. the density) should be as high as possible. The effective overburden pressure on the upper clay liner which will be exposed to low-electrolyte water, will rarely exceed about 150 kPa. As discussed previously, the swelling pressure of the clay liner material should not exceed the overburden pressure if heave of the surface is to be avoided. Using the FIM clay as an example, this means that the density of the water-saturated smectite-rich clay should not exceed 1800–1900 kg/m^3.

For the purpose of this discussion, we will assume once again that the bottom liner does not have any overlying synthetic membrane (HDPE) or LCS, that is, the clay layer is the only 'impermeable' liner underlying the waste. Continuing with the FIM clay example, if we assume the bottom liner to have an effective overburden pressure of 200–300 kPa, the corresponding maximum density at fluid saturation is about 1900–2000 kg/m^3 – considering permeation by a waste leachate rich in salts. While the hydraulic conductivity of the upper clay liner will be about E–10 m/s, the bottom liner may have a conductivity that is 5–100 times higher. The transmissivity is significantly higher than that of the upper clay liner, and since the hydraulic gradient may be much higher, the flux is also much higher. Reminding ourselves that the clay liner is the only impermeable liner underlying the wastepile, mass conservation tells us that if the flow of water is to be continuous in the system, the flux through the bottom liner cannot exceed that through the top liner. This means that the top liner will control the entire percolation process. For a single engineered clay liner system, without benefit of overlying synthetic membrane and LCS, it would appear that there is no merit in constructing a bottom liner that is tighter than the top layer. This leads to the conclusion that a greater effort should be put into constructing a very tight top liner. For the ideal case with no penetration of water through the upper clay layer, since water is the carrier for contaminants, mass balance analysis tells us that one should not expect contaminants to escape through the bottom layer into the biosphere [5].

Transport of chemical species through the bottom liner

The cation exchange capacity (CEC) of a bottom clay liner that is rich in Na smectite is about 100 meq/100 g, and about 40 meq/100 g for FIM clay. The initially sorbed Na will be replaced by cations released from the waste in the order: $Li^+ < Na^+ < Cd^{2+} < K^+ < Ca^{2+} < NH_4^+ < Co^{2+} < Cu^{2+} < Hg^{2+} < Fe^{3+}$. For typical Swedish ash from incinerated organic material, the average concentration of heavy metals (e.g. cadmium) in the porewater reaching the lower clay liner can be as high as 6 meq per litre. Most of the heavy metals will be partitioned onto the clay-liner soil solids.

The amount of water that passes through a 1 m thick bottom liner in a certain period of time is the product of the average flow rate, the cross-section area and elapsed time. We use again the liner of FIM clay as an example and assume that the liner is 1 m thick, with a density of 1950 kg/m³ and a hydraulic conductivity of K = E–10 m/s. For a hydraulic gradient of 20, and for a fully fluid-saturated wastepile of 20 m height, the annual percolated amount of water will be 60 litres. This will give us 360 meq of heavy metals in the 60 litres. Assuming that sorption of the heavy metals is equal to the cation exchange capacity (CEC) of the underlying soil, as a very simple calculation model, we will obtain (in theory) 9 kg of dispersed clay particles as a requirement for total heavy metals sorption. In actual fact, the natural soil system will not be composed of dispersed individual particles that will each sorb the contaminants. Nevertheless, the simple calculation model provides us with a 'scoping-type' calculation that can assist in the assessment of the potential capabilities of candidate materials for the liner. Laboratory leaching column tests using the candidate material and a representative leachate will provide one with a measure of the amount of correction needed for the simple calculation procedure. Carrying forward the simple model, we see that for a bottom clay liner dry density of 1500 kg/m³ corresponding to a bulk density of 1950 kg/m³ at fluid saturation, the sorption (carrying) capacity will be exhausted in a period somewhat less than 200 years. Transport of metal cations will proceed at a constant rate once the carrying capacity is exhausted [5]. This means that the fugitive heavy metal ions (ions escaping from the landfill through the saturated clay liner) would be around 360 meq of metal ions per year. If we assume that all the metal ions are cadmium ions, this would correspond to an annual escape in the amount of 42 g/m² (horizontal base area of the lower clay liner). If we: (a) take into account the flow-control exercised by the upper liner and (b) assume that the rate of water percolating through the upper liner is 1/60 of that passing through the bottom liner, the calculations will show that the elapsed time period before significant amounts of fugitive metal ions appear below the bottom liner will be some thousands of years.

When diffusion, as a transport mechanism, is factored into the calculations, the entire scenario changes because water saturation of the bottom liner is not a major factor. Diffusion transport does not require full saturation. It only requires that water is available for diffusive transport. This is generally easily satisfied by the presence of hydration water surrounding the particles. Using some common diffusion coefficients, that is, D = E–10 to E–9 m²/s [9], one finds that in the absence of an overlying synthetic membrane and LCS, heavy metals will appear below the bottom of the clay liner in low concentrations a few years after generation of the waste leachate within the wastepile. The significant difference in time for the appearance of the fugitive metal ions below the bottom clay liner reinforces the point that although water is the carrier for transport of contaminants, there is a vast difference between transport by advective means and by diffusive mechanisms. Advective transport requires near-saturation of the liner whereas diffusive transport only requires continuity in water film boundaries throughout the thickness of the liner. Since previous calculations have shown that complete saturation of the bottom clay liner will require many thousands of years, it is not surprising that calculations for advective transport of the fugitive metals would also require that many thousands of years.

Complete saturation of the wastepile (waste mass) is not needed as long as the generated leachate can flow within the wastepile and enter the bottom-liner system. This will set up the opportunity for diffusion transport of the metal ions through the bottom-liner system. Defence against early appearance of fugitive contaminant ions below the bottom-liner system is obtained with double-membrane liner systems and an LCS as shown in Figures 6.2 and 6.3. Appearance of the fugitive heavy metals below the clay liner will only occur after failure of the HDPE overlying the clay liner.

In addition to the impact of aggregation of particles forming microstructural units (MUs), we have the influence of pH on sorption, and the impact of inorganic and even organic ligands in the leachate stream competing for the heavy metals. All of these factors make it difficult to offer exact computational predictions for the required amount of clay soil material. Accordingly, scoping calculations using the simple dispersed particle model is one way of estimating material types and critical times and dimensions.

Another method for estimating the thickness required and time for full utilization of the carrying capacity of the clay liner is to conduct laboratory leaching tests. Column leaching tests using actual compact clay liner samples and actual leachates can be conducted to provide experimental information on the sorption characteristics of the candidate soil. Figure 6.19 shows column leaching test results for three soils. The profiles for concentration of metals retained in the soil in relation to depth for Pb, Cu and Zn are given

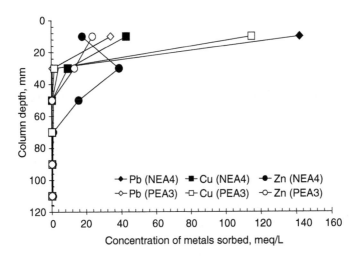

Figure 6.19 Retention profiles for Pb, Cu and Zn for estuarine alluvium samples Neath (NEA) and Newport (PEA). Retention values are obtained from acid digestion of samples subject to five pore volumes leaching column tests with a municipal solid waste leachate spiked with Pb, Cu and Zn. (Data from Yong *et al.* [14].)

in terms of meq/L. This is because determination of the metals concentrations was obtained from acid digestion of the samples at the various locations along the leaching column. The samples tested were estuarine alluvium soils identified as Neath (NEA) and Newport (PEA). The PEA soil has 16.5% carbonates, 13.4% oxides and 3.8% organics. In contrast, the NEA soil had 3% carbonates, 9.9% oxides and 5.1% soil organics. The specific surface area (SSA) for both soils are almost similar, that is, 75 m^2/g for PEA and 73.3 m^2/g for NEA. However, because of the higher proportions of carbonates and oxides, the CEC for the PEA was more than twice the NEA value (39.4 compared to 14.8 meq/100 g respectively). We see that even though the two soils are classed as estuarine alluvium soils, the functional groups of the different soil fractions combined with their influence on the development of soil microstructure contributed to the differences noted.

Impact of exogenic processes – a matter for safety analysis

Top liners   The top liners are exposed to weathering forces in the form of wind, heat, rain and frost. Some simple resultant effects include soil freezing and frost heaving, desiccation and dust production, gas penetration, erosion and piping. Freezing causes local ice-lensing to occur. When both water availability and a frost susceptible soil condition are obtained, frost heaving will result. At this stage, the ice lenses that are responsible for the frost heave are significantly larger than the local ice lenses formed in the general freezing process.

The problems arising from local ice-lensing and particularly from frost heave are not only the heave of the top surface in winter, but more so the porewater in the large cavities of the soil when the ice lenses melt (thaw). If drainage is constrained, excess porewater pressure (i.e. porewater overpressure) will develop, resulting in a decrease in the effective stress in the material. It is eminently clear that the clay liner material will be less homogeneous and that local weak spots will replace the space formerly occupied by the ice lenses. The obvious solution to the less than desirable clay liner degradation from frost effects would be to locate the liner material below the frost line. This may require the top liner to be covered by 5 m of overburden in northern Sweden, Finland, Russia and Canada. If the situation does not permit such an overburden cover, it then becomes all the more important to design the clay liner such that frost heave is denied. This means denying access to water and a soil that will not frost heave. Much has been written about the frost heave phenomenon and a considerable body of information can be found concerning the gradation of soil particles required to be frost heave free.

Desiccation can cause fracturing if the water content drops below the shrinkage limit (Atterberg consistency limit). Whereas the water retention potential of smectite clays is higher in comparison to other clay types, their water contents at complete saturation are correspondingly higher. The risk of cracking will also be higher. The risk can be reduced by decreasing the smectite content. However, this will raise the hydraulic conductivity and make the clay less self-sealing. The best way of avoiding desiccation is to locate the clay layer beyond the reach of roots of vegetation and at a depth where temperature is low and stable.

Gas penetration will cause channels as described in Chapter 5. These can serve as quick pathways for water unless they do not self-seal after peristaltic gas penetration. This may occur over long time periods. Self-repair is estimated to require a smectite content of at least 30%.

As described in Chapter 5, piping of clay liners can occur if the hydraulic gradient reaches a critical value. Extensive laboratory tests [9, 15] indicate that this is in the order of 20–30 m/m. Such high gradients will not materialize if the clay layer has a thickness of more than a few decimetres. For thin layers and some GMs, piping may represent an important problem.

Bottom liners

The carrying capacity of the clay soil in the bottom clay liner with respect to contaminant transport refers to the capacity of the soil to partition (sorb onto the soil solids) the contaminants in the transport process. When the carrying capacity is reached, the soil is considered to have sorbed its full load of contaminants. The soil is no longer capable to function as a sorption barrier. The carrying capacity of the clay soil in the bottom liner will not be exceeded until after several tens of years or even centuries. Even after this, sorption has a retarding impact on the part of the diffusive transport that is related to surface diffusion. However, pore diffusion, which makes up about 30–50% of the total diffusive flux depending on the bulk density of the clay [16], will remain constant. Depending on the pH of the microenvironment and the presence or absence of inorganic and organic ligands, a portion of the heavy metals will be precipitated and another portion will form complexes with the ligands and be transported through the soil.

Concluding remarks

The basic principle articulated in this chapter is with respect to minimization of contaminant transport through and out of the wastepile. In simple terms, this means covering the wastepile with very tight upper clay layers with a moderate smectite content. Under normal conditions, no special precautions have to be taken with respect to chemical degradation of these layers or to piping. It is, however, necessary to protect them (the top layers) from erosion and freezing. Bottom clay liners are not solely responsible for the rate of flux of water through the waste pile: if they are tighter than the upper clay liners they will, for a period of time, let less fluid through than the upper clay liner. However, after fluid saturation of the entire system,

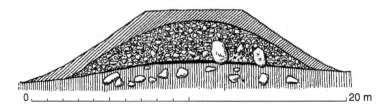

Figure 6.20 Three thousand year old mound at Hersby, Sollentuna, Uppland in Sweden (After H. Lindqvist).

electrolytes will change this and the upper clay liners will play a dominant role in the control of the flux.

Historical evidence suggests very strongly that in locations where the hydrogeological settings are favourable, disposal of waste in hill-type landfills located above the regional ground surface should not require collection and treatment of leachates until at least several thousand years – provided that the landfill has been properly constructed. This is validated by the dry conditions in 3000 year mounds from the Bronze Age (1500–1100 BC) such as the Haaga Hoeg (King Bjoern's grave) in Uppsala not far from Stockholm. This and a number of similar barrows in Denmark and Germany have a thick clayey cover of rock block structures with joints sealed with clayey soil. Very well preserved oak coffins have been found in the interior of the dry block structures. Figure 6.20 shows a schematic section of a small mound in the Stockholm area. In Germany, the diameter of such mounds are up to 90 m and the height up to 15 m, yielding a slope angle of about 18°, that is, in the same order of magnitude that we calculated in Section 6.2.4. This suggests that our ancestors were very proficient in soil mechanics.

6.3   Clay for isolation of highly radioactive waste – deep geological disposal

6.3.1   *Processes in the near-field of heat-producing waste*

The major difference between disposal of chemical waste and high level radioactive waste (HLW) is that the latter produces heat in conjunction with radioactive decay and gamma radiation. This means that this type of waste be it processed uranium fuel from nuclear reactors or unprocessed spent fuel has to be kept inside canisters, which are commonly made of iron, steel or copper. According to most concepts proposed by the organizations that are responsible for handling and disposal of HLW, the canisters will be placed in holes or tunnels bored at a depth of several hundred metres in crystalline rock, salt, argillaceous rock or plastic clay.

The heat produced will affect the rock and its isolation potential, making it necessary to separate it from the hot canisters by a suitable material. Clays have been proposed as the buffer material because it can provide:

- very tight embedment of the canisters to minimize water flow around them;
- the necessary properties to retard transport of corrosion-promoting ions from the rock and migration of ultimately released radionuclides;
- ductile embedment of the canisters to smooth rock stresses caused by tectonics and rock fall.

These functions require that the clay buffer: (a) becomes water-saturated soon after placement to provide effective heat transport to the surrounding rock and (b) to expand, so as to provide perfect contact with the canisters and rock. The expandability and self-sealing potential must be robust, that is, these properties must continue to exist for periods as long as tens to hundreds of thousands of years. Effective isolation in the face of potential rock and canisters movements must be achieved. Self-healing capability is required because these movements can create heterogeneities, and also because initially formed desiccation fractures must be healed. All these functions require that mineralogical changes should be relatively insignificant in this long time period and that at least 50% of the smectite remains unaltered after the required isolation time. The early evolution of the buffer is important, particularly with respect to the wetting rate. However, the impact of temperature in the several hundred or thousand year long hydrothermal period is believed to be even more important for survival of the buffer.

The Swedish concept KBS-3V for underground storage of HLW, which is one of many concepts for HLW isolation, is used here as an example to focus the discussion. In this concept the canisters are embedded in blocks of highly compacted Na smectite, such as MX-80, and placed in bored deposition holes with 1.8 m diameter and 8 m depth, in crystalline rock at about 500 m depth (Figure 6.21). The deposition holes have a spacing of about 6 m and the canisters produce energy that is about 1000 W at the time of placement. It is estimated that the time for heat decay to reach final temperature equilibrium within the repository will be about 3000 years. The maximum temperature at the canister surface will be about 95°C and the temperature gradient across the buffer will become vanishingly small after about 1000 years.

The early maturation stage

The following changes with respect to temperature (T), hydraulic conditions (H), mechanical, that is, stress/strain (M), chemical (C), biological (B) and radiological (R) conditions will occur in the system of highly compacted

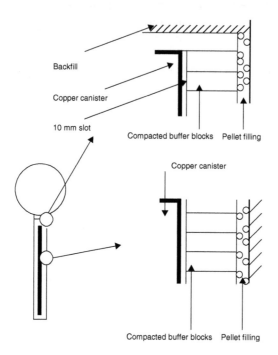

Backfill

Copper canister

10 mm slot

Compacted buffer blocks   Pellet filling

Copper canister

Compacted buffer blocks   Pellet filling

Figure 6.21 KBS-3V deposition hole with copper-shielded canister embedded in compacted buffer blocks and bentonite pellet filling.

smectitic blocks (MX-80), confined by the overlying backfill and the surrounding rock, as indicated in Figure 6.22:

- Thermally induced redistribution of the initial porewater results in desiccation in the hottest part of the buffer and wetting of the colder, outer part of the buffer.
- Maturation of the pellet fill leads to water saturation, homogenization and consolidation of the buffer component under the swelling pressure exerted by the hydrating and expanding blocks.
- Uptake of water from the rock and backfill leads to hydration of the buffer.
- Expansion of the buffer, leads to an eventual tight contact between canister and buffer and rock. However, this will also cause upward displacement of the canisters and the overlying backfill.
- Chemical processes occur within the buffer and at the contact of buffer and canisters. A most important process is that salt, primarily Ca, Cl and SO_4, will migrate with the water that migrates from the rock. Precipitation in the hot part of the buffer will produce solid or brine NaCl and gypsum, contributing thereby to corrosion of the canisters.

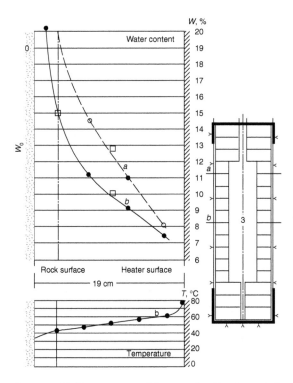

Figure 6.22 Temperature-driven redistribution of the original water content (10%) of very dense MX-80 clay surrounding a 600 W heater in the Stripa field experiment (Buffer Mass Test) after a few months. The rock provided very little water to the buffer [15].

The various processes, which are coupled in a complex way will cause transient changes in the material properties. This makes analytical-computer modelling of system performance and buffer performance difficult. To render the problem more tractable, a number of simplifications have been made in the various prototype repository performance assessment exercises. One of the simplifying assumptions concerns the buffer material. The smectitic buffer is considered to be a homogeneous medium, although it is in actual fact highly heterogeneous, with water vapour and water moving in the different types of void spaces.

The pellets in the 50 mm water-filled gap between the smectite blocks and the rock face cause difficulties in thermal–hydraulic–mechanical– chemical–biological (THMCB) modelling. This is due to the difficulties in producing a conceptual model that reflects the processes involved in hydration and subsequent performance of the pellets. The pellets disintegrate quickly and form a soft clay gel, allowing water to be sucked by the very dense clay blocks that form the buffer. Expansion of the clay blocks will consolidate the gel, resulting in a clay medium that will be as dense as the expanded blocks. The process has a B-effect – meaning that microbes can enter the pellet filling and survive and multiply until the density of the filling reaches about 1600 kg/m^3. The transient conditions are illustrated as follows:

- The hydraulic conductivity of the outermost part of the buffer drops and the inflow of water from the rock into the deposition holes decreases.
- There is competition with respect to water between the densifying pellet filling and the softening buffer blocks.
- The consolidated clay gel formed by the initial isotropic wetted pellet fill becomes anisotropic due to the vertical particle orientation induced by the high horizontal pressure.

Water saturation may require a decade for a buffer in richly water-bearing crystalline rock, several decades in tight rocks of this type, and hundreds to thousands of years in argillaceous rock. All the physico/chemical processes described heretofore occur during this long period, and may lead to significant changes especially in the early phase when the temperature gradient is high. The ability of the surrounding rock to supply the buffer with water is therefore a key question in the design and performance assessment of the repository concepts.

6.3.2   Maturation of buffer clay

The hydrothermal period

After complete water saturation, the buffer enters a period of several hundred to thousand years where hydrothermal conditions prevail. The changes resulting from this condition will be described in detail in Chapter 7. In short, the smectite degrades by dissolution and undergoes alteration to non-expandable minerals such as illite and chlorite. In consequence, the clay buffer tends to become brittle due to cementation caused by precipitation of silica released from the dissolved minerals. Cementation can also result from chemical interaction between canisters and buffer, that is, by formation of iron compounds. Some of these changes are slow, however, and some may be reversible when the temperature gradient ultimately vanishes, and may not threaten the overall long-term performance of the buffer. There is a high possibility that, in fact, precipitation of silica may have a sealing effect on the surrounding rock. Gamma radiation will have only a very small impact on the physical and chemical states of smectite minerals. This has been documented by exposing clay samples (MX-80) to strong gamma radiation under hydrothermal conditions [15].

*Impact on the evolution of buffers by
the surrounding rock*

The performance of the surrounding rock on the clay buffer is a critical issue. The major factors to be considered with respect to rock performance are: (a) the ability of the host rock to allow water to reach the buffer for early maturation and (b) its capability to allow gas to escape from the buffer. The latter issue is of fundamental importance because very high pressures generated by gas build-up without release may result in the crushing of both near-field and far-field rock – as well as of canisters. The problem is minor for crystalline rock, but may be a critical factor in the disposal concepts for salt and argillaceous rock.

Figure 6.23 illustrates the function of the near-field rock around a deposition hole for canisters. The most important role for the rock is its influence on the maturation and performance of the buffer. The rock sets the boundary conditions for water saturation in the early maturation phase similar to the conditions for liners of waste landfills. The hydraulic transport capacity of the rock surrounding deposition holes or tunnels depends on the frequency

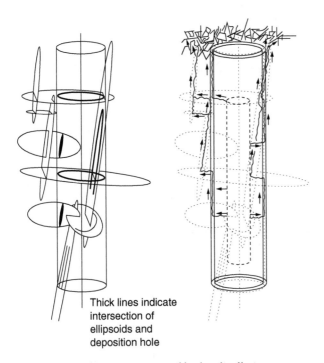

Thick lines indicate
intersection of
ellipsoids and
deposition hole

Figure 6.23 Schematic picture of hydraulically important structural features in near-field rock. Left: system of interconnected discontinuities with varying water-bearing capacity. Right: transport paths of gas and radionuclides from leaching canisters up to the floor of blasted tunnel via the continuous boring-disturbed zone [15].

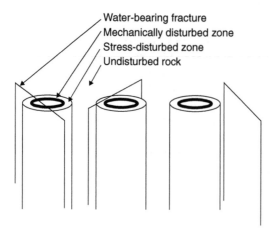

Figure 6.24 Different hydraulic boundary conditions represented by cases with water-bearing fractures intersecting a waste-deposition hole in the rock, fractures intersecting the EDZ around the hole, and conditions with no such fractures. The access of water for buffer maturation can vary by several orders of magnitude.

and conductive properties of the discontinuities in the rock. The distribution of these discontinuities over the periphery of the holes is controlled both by the location of intersecting fractures and by the conductivity of the shallow boring-disturbed zone (excavation disturbed zone (EDZ)).

The frequency of water-bearing fractures in crystalline rock, in respect to desired quality site location and design stages of a HLW repository, is relatively low in crystalline rock and extremely low in argillaceous rock. In deep salt domes and bedded salt, they do not appear at all. The typical structural constitution of crystalline rock illustrated in Figure 6.24, shows that an 8 m deep deposition hole is intersected by only 2–3 water-bearing fractures. They are the ones that control the access to water for the maturation of the buffer. The very low frequency of such features in argillaceous clay is responsible for the long time period required for water saturation of the buffer, a condition that is favourable for physico/chemical processes (described earlier) in the buffer material.

The EDZ of bored holes is caused both by excavation damage forming new fractures, and by propagation, opening or closure of existing ones depending on the prevailing stress conditions. For argillaceous rock, the EDZ is the only rock structural component of practical importance for water and gas transport. In the case of crystalline rock, long-extending fractures are most important. The shallowest part of the walls of a deposition hole is damaged by the shear stresses induced by the rock boring process. They have the form of partly broken crystal bonds and very fine fissures of the type shown in Figure 6.25. This figure shows disturbance in the form of

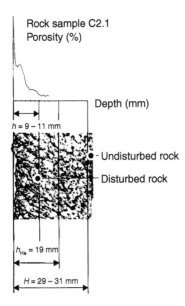

Rock sample C2.1
Porosity (%)

Depth (mm)

h = 9 – 11 mm

• - Undisturbed rock

— Disturbed rock

h$_{He}$ = 19 mm

H = 29 – 31 mm

Figure 6.25 Interpretation of thin section of crystalline rock with respect to mechanical disturbance by TBM-type boring (after J. Autio).

breaks ranging from microscopic to about 10 mm depth from the free surface. The average porosity is at least 10 times higher within 3 mm distance from the surface in comparison to undisturbed rock, and 2–5 times higher in the interval 0–10 mm from the surface.

A shallow boring-disturbed zone is beneficial to the wetting of the buffer clay because: (a) it has a higher hydraulic conductivity than the crystal matrix of the surrounding rock and (b) of its capability to distribute water coming in from discrete fractures over the periphery of deposition holes and tunnels. Discussion of water redistribution within the buffer under the thermal gradient existent in the first hundreds or thousands of years in Chapter 5, focussed on the basic processes responsible for driving water in the saturating clay. Although it is generally believed that the driving forces for wetting of the buffer are the internal gradients (manifested as suction), the detailed mechanisms in water transport have yet to be fully developed. Microstructural examination suggests that hydration takes place by two parallel mechanisms: migration along particle surfaces from the wet outer boundary and condensation of vapour-transported water from the hot inner part of the buffer. Since both processes are of diffusion-type, one can describe buffer saturation as a quasi-diffusive process. Kröhn and others [17] have claimed good agreement between calculated results obtained from modelling and actual measured values of vapour diffusion under isothermal conditions – using actual determinations of the water content as

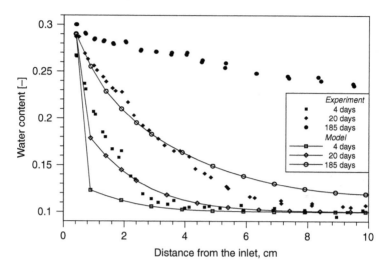

Figure 6.26 Water content as a function of the distance form the wet boundary [17].

illustrated by Figure 6.26. They conclude that if vapour diffusion fades to Knudsen-diffusion in an advanced stage of saturation the buffer, water migration can be explained exclusively by vapour flow due to an increased diffusion coefficient. There are many examples that one can consider in the wetting of smectitic soil where the wetting occurs as a diffusive process. Chapter 8 deals with the modelling of important processes in engineered clay-based barriers.

Boundary conditions

Application of any model for prediction of wetting of buffer clay requires specification of the hydraulic boundary conditions. The three most important parameters to be used in applying the THM-models are (1) the piezometric pressure in the near-field rock, (2) the frequency of hydraulically active fractures that can bring water to the deposition holes, representing the average hydraulic conductivity of the near-field rock and (3) the hydraulic conductivity of the EDZ.

We will illustrate the significance of the hydraulic boundary conditions using the basic case of a KBS-3V deposition hole in crystalline rock as an example. The diameter of the holes is taken as 1.75 m and the buffer column is assumed to fully occupy the holes. The clay has a dry density of 1850 kg/m^3 and the initial degree of water saturation is 50%. Saturation of the buffer occurs through processes resulting from capillary and osmotic suction. Migration of water occurs by diffusion whereas water transport in the surrounding rock is by Darcian flow. The net approximate diffusion coefficient of 3E–10 m/s² determined in laboratory tests has been found to be in agreement with earlier-performed field tests. These tests were conducted under condition of moderate temperatures and temperature gradients.

General performance

It is possible to distinguish between: (a) the case with more water moving to the deposition holes (not taken up by the buffer) and (b) the opposite case with a deficiency in water supply to the buffer. In the first-mentioned case the water pressure at the rock/buffer contact increases with time while in the opposite case, it will initially decrease and will subsequently increase. The hydraulic conductivity of the rock and the water diffusivity of the buffer combine to determine the rate of hydration of the buffer.

Diffusive transport in the buffer clay is controlled by the expression:

$$\frac{\partial C}{\partial t} = D\nabla(\nabla C) \tag{6.12}$$

where the initial value of the concentration C, which is here the degree of water saturation is 0.5.

Water flow in the rock is given by the expression:

$$S\frac{\partial \phi}{\partial t} = K\nabla(\nabla \phi) \tag{6.13}$$

where S is the storage coefficient, K is the hydraulic conductivity, ϕ is the water pressure in metre water head and t is the time.

The porosity of the clay is $P = V_{por}/V_{tot}$, which yields:

$$V_{por} = PV_{tot} \tag{6.14}$$

We can express the degree of water saturation as:

$$C(r = r_1) = \int \frac{q_w(r = r_1)}{V_{por}} dt \tag{6.15}$$

where $q_w(r = r_1)$ is the water flux at the boundary $r = r_1$ (the periphery of the hole). V_{por} is given by Eq. (6.3) t = time.

By applying simple potential theory, one gets for the flux towards the hole $q_w(r = r_1)$ the expression in Eq. (6.16), cf. Figure 6.27:

$$q_w = \frac{K2\pi b(\phi_2 - \phi_1)}{\ln(r_2/r_1)} \tag{6.16}$$

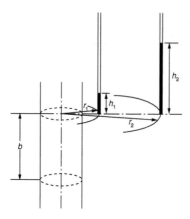

Figure 6.27 Flow conditions in porous medium with long hole.

A pressure deficit in the clay at $r = r_1$ brings water across the periphery ($r = r_1$) and since mass balance is required one gets the degree of water saturation C for $r = r_1$ as expressed by Eq. (6.17):

$$C(r = r_1) = \int \frac{K2\pi b(\phi_2 - \phi_1(t))/\ln(r_2/r_1)}{PV_{tot}} dt \qquad (6.17)$$

where P is the porosity of the clay and V_{tot} its total volume.

We assume here that the radial water pressure distribution for the first time step is obtained by solving Eq. (6.17) for the pressure value representing r_2 at the outer boundary, taken here to be 4 MPa, and 0 MPa pressure at the hole periphery (r_1). This gives $C(r = r_1)$ from Eq. (6.6) for each time step for application in Eq. (6.17).

The same procedure is necessary also for the case with a 'wet' boundary, that is, unlimited access to water, since it is not possible to know when, in the simulation of $C(r = r_1)$ given by Eq. (6.17), C is higher or equal to unity. This is accomplished by watching $C(r = r_1)$ in the course of the simulation, and specifying $C(r = r_1) = 1$ if it exceeds unity. Applying this mode of calculation, one gets the rate of increase in the degree of water saturation of the clay with time – as shown in Figures 6.28–6.30 for the three rock types.

*Influence of the hydraulic boundaries on
the wetting rate*

One finds from the example that for a rock with K = E–10 m/s, it would take more than 75 years to reach complete water saturation at the centre of the clay column. At the mid-height of the canister, where the inner clay radius is 0.45 m, the degree of water saturation would only be raised from 50% to 65% in the first 10 years. After 25 years this figure would be about 80%.

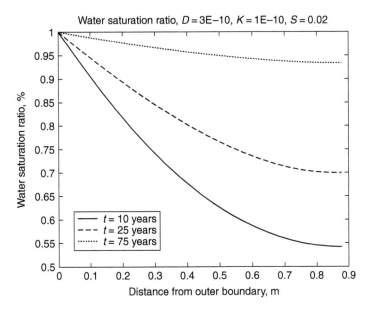

Figure 6.28 Increase in saturation of the clay for rock with K = E–10 m/s.

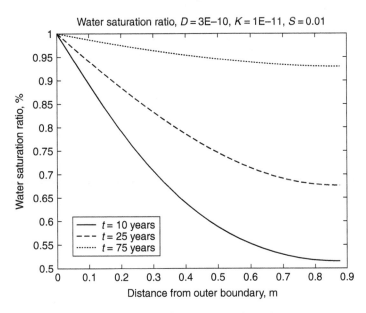

Figure 6.29 Increase in saturation of the clay for rock with K = E–11 m/s.

The most important finding is that rocks with the highest conductivities develop water pressures at the clay/rock interface higher than zero from start, and that the degree of saturation of the buffer therefore steadily increases. In contrast, there will be a deficit pressure in the water at the rock/buffer interface for a considerable period of time in the tightest rock.

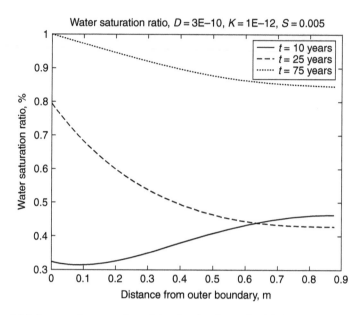

Figure 6.30 Increase in saturation of the clay for the rock with $K = E{-}12$ m/s. Lowest
curve implies initial suction of water from the buffer clay to the rock.

Figure 6.29 demonstrates that the rock will actually suck water from the
clay and that the degree of water saturation will be lower than the initial
50% value for more than 25 years. In practice this will not occur in
crystalline rock since the bulk conductivity is hardly ever lower than $E{-}11$ m/s.
In argillaceous rock, however, this condition may appear.

The case with insufficient water supply for undisturbed water uptake by
clay is valid also for top and bottom liners of waste landfills. In fact, this
state is reached numerous times in the operational lifetime of the liners,
especially for top liners. This is due to the fact that periods of drought will
be frequent, resulting not only in a cessation of the wetting process but even
an evaporation of clay-sorbed water.

6.3.3   Other applications of dense smectite clay as engineered barrier

General

The excellent sealing properties of dense smectite clay will be used for other
purposes as well in HLW repositories. Amongst such applications are:

- Backfills in drifts, tunnels and shafts.
- Plugs in drifts, tunnels and shafts.
- Plugs in boreholes.

Backfills in drifts, tunnels and shafts

The primary purpose of backfill in tunnels and shafts is to seal the long rooms and to provide support to the rock. They will serve as unwanted, effective flow paths if the backfill has a higher hydraulic conductivity than the surrounding rock. Considering that the EDZ of blasted rooms can be as high as E–8 m/s, this criterion would imply a maximum conductivity of the backfill of about E–9 m/s. This can be attained by compacting smectitic clay or mixtures of smectite and ballast to the required density. The EDZ has to be cut off by constructing tight plugs that extend sufficiently deep into the surrounding rock to create stagnant groundwater conditions.

Materials for backfilling should meet the following general requirements:

- Lower hydraulic conductivity than the surrounding rock
- Exertion of a swelling (effective) pressure that prevents rock fall
- Capability for placement even under the condition of significant water inflow
- Rational placement.

These criteria make it necessary to use smectitic material for backfilling. Accordingly, the physical properties of mixed or natural clays become important for designers and modellers of HLW disposal. The problem is to produce a backfill that is homogeneous and sufficiently dense. Various methods have been tried, including: (a) traditional road construction techniques with compaction in layers, (b) compaction of a suitable clay/ballast mixture up to a certain level in horizontal drifts and tunnels and (c) filling the remaining space by blowing in dense pellets of smectite-rich clay. The last technique (blowing in dense pellets) cannot be used if the inflow of water from the rock per metre tunnel length exceeds about 1 l/h – because the backfill will become slurry, and piping and erosion will result.

Another technique in application and compaction of inclined thin layers is shown in Figure 6.31. Discharge of inflowing water from the rock is measurably simpler than in the first-mentioned case. However, difficulties exist with respect to water inflow where the tunnel intersects fracture zones in crystalline rock. In practice, comprehensive grouting of the rock is required to reduce the risk of piping and erosion. Only a very tight lining or freezing of the rock would eliminate the problem of inflowing water. This is of little or no concern in argillaceous rock.

The problem with placement and *in-situ* compaction of backfills in a closed space is the energy loss at compaction near the perimeter. It usually results in a significantly lower backfill density within a few decimetres from the physical boundary as illustrated by Figure 6.32. This has led to the idea of backfilling with compacted blocks of clay/ballast mixtures or of suitable natural clay such as FIM clay. A masonry of blocks of this material with a net dry density of 1700 kg/m^3 will ultimately homogenize and mature with

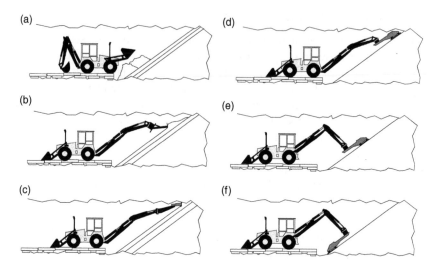

Figure 6.31 Placement (a, b, c) and compaction (d, e, f) of backfill using the 'inclined layer principle'. A special vibratory roof compactor for densifying material close to the roof is not shown [18].

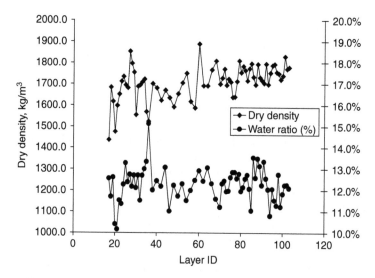

Figure 6.32 Variation in dry density and water content as functions of the number of applied layers of mixed smectite clay and suitably graded crushed rock [18]. The lowest density yields a hydraulic conductivity, for a saline porewater, of 5E–9 m/s. The highest density produces a hydraulic conductivity that is 20 times lower.

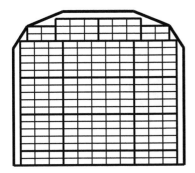

Figure 6.33 Clay block masonry with clay slurry (grout) filling the gap to the rock [1].

a density at water saturation of about 2000 kg/m³ (Figure 6.33). The net hydraulic conductivity is less than E–10 m/s even for saline porewater, and the developed swelling pressure is more than enough to support the surrounding rock ($\gg 100$ kPa). A similar recently proposed technique for applying solid chemical waste in drifts can also be used for backfilling (EC-project LowRiskDT, Contract No. EVG1-2000-00020). This may be applicable to situations where water inflow is relatively strong.

Figure 6.33 shows the basic case where only little water is given off from the rock, and where the block masonry can be left with a gap to the surrounding rock. For such a scenario, water will be absorbed spot-wise, and some flow of disintegrated clay emanating from the wetting blocks will result. The resultant variations in density and time for fluid saturation can cause minor block movements and some heterogeneity. These disadvantages can be minimized by injecting a suitable clay-based grout in the gap behind a temporary (shotcrete) shield that can resist the grout pressure. This technique can also be applied section-wise where there is a relatively high inflow of water. The grout should be so composed that it initially has a low viscosity and subsequent quick strengthening. This can be obtained for instance by mixing strongly compacted smectite pellets with low-pH cement.

The grout can be made chemically compatible with the block material and be given a dry density of 1200–1400 kg/m³. The ultimate density of the saturated backfill will be 1900–2000 kg/m³ for a filling degree of the block units of 90%, and a dry density of the blocks of around 1900 kg/m³. This gives the entire backfill a net swelling pressure of a few hundred kPa and a bulk hydraulic conductivity of less than E–10 m/s. The major function of the grout is to allow the block masonry to mature uniformly while consolidating under the swelling pressure exerted by the maturing block system.

Laboratory experiments have been undertaken to develop grouts with a fairly high density and a capability for easy pumping. Grouts based on FIM

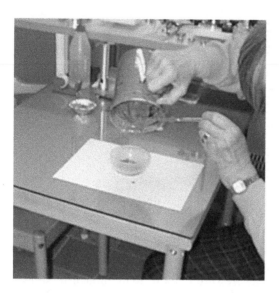

Figure 6.34 Freshly prepared mixture of FIM clay, slag cement and strongly compacted MX-80 pellets. The mixture had a density of about 1200 kg/m^3 and was fluid but stiffened considerably in minutes.

clay mixed with 5% slag cement and strongly compacted MX-80 pellets and with a water content of about 120% are sufficiently low-viscous materials. They can be pumped (Figure 6.34) and will stiffen fairly rapidly. They exhibit an increase in shear strength from 1 to 10 kPa in one day and to 100 kPa in a week.

When applied on site, the grout in the gap between rock and compacted blocks will consolidate and stiffen even quicker. It is expected to resist piping if the pressure of water inflowing from rock fractures is not too high. Systematic tests with grouts of different compositions are required to assess the potential value of the method. Tests with FIM clay in contact with saturated cement water have indicated considerable resistance to chemical changes of the clay.

Backfills for embedment of large structures

A related application of backfills is illustrated by the clay-based engineered barriers used for isolating a concrete silo at 100 m depth in crystalline rock for hosting intermediate-level radioactive waste (Figure 6.35).

The bottom bed of the silo is constructed so as to yield minimum compression under the pressure exerted by the 16 000 ton silo, and to have a hydraulic conductivity of no more than E–10 m/s. The first-mentioned criterion requires a low clay content and a high density of the mixture.

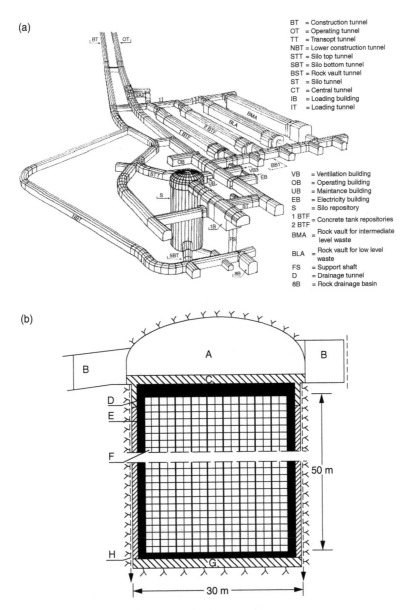

(a)

BT = Construction tunnel
OT = Operating tunnel
TT = Transopt tunnel
NBT = Lower construction tunnel
STT = Silo top tunnel
SBT = Silo bottom tunnel
BST = Rock vault tunnel
ST = Silo tunnel
CT = Central tunnel
IB = Loading building
IT = Loading tunnel

VB = Ventilation building
OB = Operating building
UB = Maintance building
EB = Electricity building
S = Silo repository
1 BTF = Concrete tank repositories
2 BTF
BMA = Rock vault for intermediate level waste
BLA = Rock vault for low level waste
FS = Support shaft
D = Drainage tunnel
8B = Rock drainage basin

(b)

Figure 6.35 Repository consisting of a big concrete silo for intermediate level waste, and vaults for low-level waste at Forsmark, Sweden. Lower: Schematic cross-section of the silo cavern. (A) Cement-stabilized sand, (B) Concrete plugs, (C) Bentonite/sand top bed with gas outlets, (D) Concrete, (E) Bentonite granules, (F) Waste, (G) Bentonite/sand bed, (H) Drains connected to tunnel system [15].

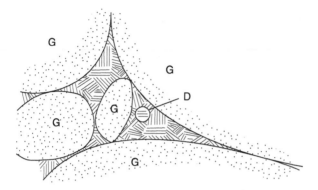

Figure 6.36 Schematic picture of the microstructure of a clay/ballast mixture
with low compressibility and low hydraulic conductivity. G =
ballast grains, D = clay powder grains [9].

The second criterion requires the use of a sufficiently high clay content to
fill the ballast voids effectively. This leads to the selection of a clay content
of 10% and application of a very effective compaction technique. The basic
principle is to obtain a microstructure consisting of contacting ballast
grains that transfers stresses, and a dense clay component filling up the very
small voids that remains after compacting the mixture to a dry density of
nearly 2200 kg/m^3 (Figure 6.36).

Plugging of drifts, tunnels and boreholes

General Radionuclides will migrate by diffusion through the buffer and
will reach the 'sink' represented by the boring-induced EDZ around the
deposition holes. From this point, they will proceed via discrete fractures to
larger water-bearing fracture zones that can bring them to the biosphere.
The process may be partly driven by pressurized hydrogen gas emanating
from corrosion of iron or steel canisters. This migration pattern has set the
basis for the international R&D priority lists for research on near-field
functions as follows:

1 The integrity of the canisters is of primary importance. It is affected by
   chemical interaction with the buffer and groundwater and by tectonic
   events.
2 The physical stability of the buffer is the second most important issue.
   Its stability and performance depends on chemical interaction with the
   groundwater and the canister.
3 The hydraulic performance of the EDZ around boreholes and tunnels
   is the third most important issue. It depends on the rock structure and
   excavation technique, both of which require more systematic study.

4   The physical stability of the backfill is of importance as well, since compaction under its own weight and chemically induced increase in porosity will provide quicker transport of water, gas and radionuclides. Liquefaction of its upper part may occur under severe earthquake action if the density is low [13].

Tunnel plugging   Plugs in tunnels will be required for many purposes. A very important application is the sealing of deposition tunnels in repositories for radioactive or hazardous chemical waste. These plugs usually have to resist high pressure from water or clayey backfill and therefore need to be keyed into the surrounding rock. In HLW repositories plugs must be tight and have to cut off the EDZ in the operational phase. Simple temporary plugs made of shotcrete may be used in the construction phase. However, where pressures from water and swelling backfills are high, plugs have to be made by casting concrete with reinforcement placed in the form. Water leakage along the plug/rock contact can be minimized by using 'O-ring'-type seals of smectite clay in the deepest part of the notch. This can be produced by slot-drilling or careful blasting [15]. Grouting of adjacent rock with low-pH cement after the construction improves the tightness. Plugs in deposition tunnels that are expected to serve even after thousands of years, can be made as masonries of compacted smectite blocks replacing the concrete plugs, in conjunction with closure of the respective tunnel system [15]. The chemical compatibility of clay and cement is the key question. This will be discussed in Chapter 7.

Figure 6.37 shows a cross-section of a concrete plug with clay seals extending into recesses to cut off the EDZ of the blasted tunnel. Figure 6.38 illustrates the practical construction. The tightness of the plug was tested by measuring the water moving through the not completely tight concrete and the surrounding rock under 500 kPa controlled water pressure on the plug. Since maturation of the clay to provide tightness took several months, the leakage dropped with time as shown by Figure 6.39.

The performance of clay seals depends on a number of microstructurally related issues. These include: (a) homogeneity and density; these will determine the risk of piping and erosion under the very high hydraulic gradient that prevails and (b) physico/chemical interactions with the concrete; these affect the various exchange processes and the chemical stability of the clay.

Borehole plugging   Early studies by organizations responsible for developing techniques for safe disposal of radioactive waste showed that boreholes extending from the ground surface down to and through a HLW repository may provide a quick and direct access for errant radionuclide to the biosphere. These studies highlighted the need for a sealant for the borehole that would retard migration of radionuclides. Research and development of methods for long-term sealing of deep boreholes have shown that smectite clay is the best candidate to fulfil the sealant role. The

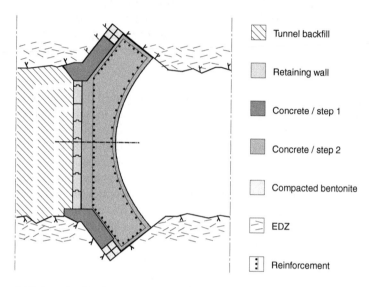

▨	Tunnel backfill
▨	Retaining wall
▨	Concrete / step 1
▨	Concrete / step 2
▨	Compacted bentonite
▨	EDZ
▨	Reinforcement

Figure 6.37 Longitudinal cross-section of rotationally symmetric plug with clay seals [19].

Figure 6.38 Geometry and appearance of the 1600 bentonite blocks ($\rho_d = 1600$ kg/m^3). Mineral wool was used for preventing cement slurry flowing into the narrow space between the blocks and the rock [19].

two candidate procedures being considered include: (1) placement of dense, precompacted blocks of smectite-rich clay and (2) filling compacted smectite pellets in holes containing drilling or deployment mud, or in dry holes with and without subsequent injection of clay slurry. In both cases,

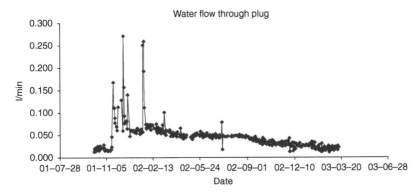

Figure 6.39 Recorded outflow through and along the plug under a water pressure of 500 kPa in the drift [19].

the plugs will tend to become homogeneous with the degree of homogeneity dependent on the density of the clay and the chemical composition of the fluid, as well as on time. Recent studies have shown that only the first method is suitable for holes deeper than 100 m.

For practical reasons, and to ensure that the plugs are located exactly as planned, the blocks are confined in metal nets or perforated tubes as described in Chapter 4. Axial confinement is required to prevent expansion in the direction of the hole. Cement can be used, provided that it is chemically compatible with the clay (low-pH cement). Boreholes with a length of about 100 m have been successfully plugged, using this technique [15]. The problem is that the placement of borehole plug segments in very deep boreholes takes time and can be compromised by untimely quick maturation of the clay component (Figure 6.40).

6.3.4 Microstructural processes in maturing clay seals

As we have discussed in the previous sections and chapters, the sealing capability of smectite clay in all the different clay-based engineered barriers is due to its very low hydraulic conductivity and significant swelling capability. The performance of dense, smectite-rich clay in borehole plugs shown in Figures 6.40 and 6.41 is a very good illustration of the volume–change swelling (expandability). This is the key issue in sealing repositories, and is a necessary property for self-sealing of voids both within the clay mass, and the boundaries surrounding the clay mass. The rate of expansion and associated change in physical properties of clay seals will be discussed in Chapter 8 when we discuss theoretical modelling procedures. We will confine ourselves here to an examination of the processes involved in the expansion of any element of smectite-rich clay into an open space filled with water. The

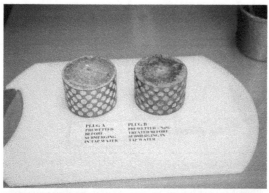

Figure 6.40 Tests of plugs of smectite clay with a dry density of 1600 kg/m³ in perforated copper tubes before and 24 hours after submersion in low-electrolyte water. (A) Without pretreatment, (B) With coating of silica solution – attempting to seal off water temporarily. After 24 hours most of the clay columns in Plug A protruded by about 6 mm and by 8–10 mm in Plug B. A very soft clay gel reaching out to about 10 mm from tube B tended to flow down along the tube, indicating exceedingly fast wetting and expansion. The reason for the malfunction was that the salt content of the silica solution (Na and K) produced rapid dispersion of the clay.

borehole plug and the case of buffer expanding into and sealing off open fractures in the surrounding rock are shown in Figures 6.40 and 6.41. Figure 6.42 shows the predicted expansionn rate.

6.3.5   Use of soft smectite clay for sealing rock fractures

Properties of soft smectite clay

If smectite clay can be introduced into rock fractures, it can effectively seal them. This has been demonstrated in various laboratory and field tests.

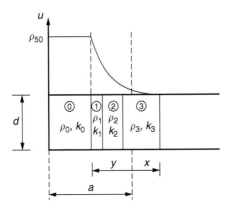

Figure 6.41 Simplified description of the process leading to expansion of clay with density ρ_0 and conductivity k_0 into a water-filled space, like fractures, gap between dense smectite clay and a boundary and penetration into the voids between ballast grains. *a* is the size of the original clay element that expands the distance *x* thereby causing changes in density (ρ_1 and ρ_2) and hydraulic conductivity. The driving force is the porewater tension *u* that drops with a reduction in density.

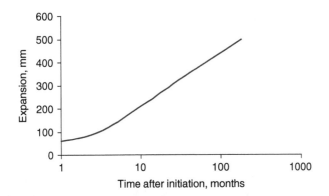

Figure 6.42 Theoretical time-dependent movement of the front of the clay element in Figure 6.41 predicted on the basis of laboratory data of the controlling parameters but neglecting the retarding effect of friction along boundaries.

The problem is that the viscosity is high if the density is high, and only soft gels can enter fractures to an appreciable depth by pressure injection. A possible solution to this would be to include electrolytes, especially Ca, into the mix in proportions that would result in a grout that would be homogeneous whilst being able to avoid coagulation. Such kinds of grouts can be effective seals as illustrated by the fact that the hydraulic conductivity is as low as E–10 to E–9 in low-electrolyte water even for gel densities of 1100–1200 kg/m^3.

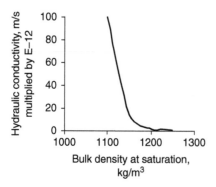

Figure 6.43 Hydraulic conductivity of soft Na-montmorillonite clay percolated with very electrolyte-poor water.

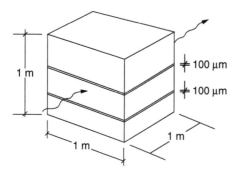

Figure 6.44 One cubic meter block with two permeable 100 μm fractures.

These densities approximately represent the upper limit of injectable clay grouts (Figure 6.43).

The net effect on the bulk hydraulic conductivity of rock by complete filling of water-bearing fractures is illustrated by the example shown in Figure 6.44 with a rock block intersected by two plane fractures with different internal geometries.

Using conventional flow theory, that is, the cube law, flow in the indicated direction gives the rock block the average hydraulic conductivity of 2E–7 to 2E–6 m/s. This will drop to a small fraction of this value if the fractures can be filled with clay cf. Table 6.2.

In practice, there are difficulties in introducing clay so that the fractures become completely filled with grout. However, once there, they stiffen thixotropically as described in Section 5.7.5. If the density is sufficiently high, that is, more than about 1100 kg/m³, the resistance against piping and erosion can be significant if the water pressure gradient across the grout depth in the flow direction is small. Experience tells us that a pressure drop

Table 6.2 Net average hydraulic conductivity K of rock block with two identical 100 μm fractures with different geometry sealed by clay grout with various conductivities

Fractures	K of ungrouted rock, m/s	K of grouted rock, m/s		
		$K_g = E{-}4$	$K_g = E{-}6$	$K_g = E{-}8$
Two plane slots	2E–6	2E–8	2E–10	2E–12
Two fractures with ten 0.01 m channels	2E–7	2E–9	2E–11	2E–13

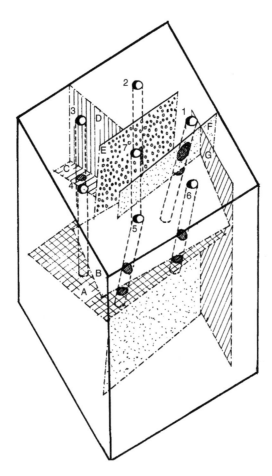

Figure 6.45 Experiment for electrophoretic sealing of fractures in a 1 m³ block of granite with five boreholes equipped with cathodes in smectite clay slurry, and a central hole to which the anode was submerged in water. A DC potential of about 1 V/cm forced clay particle aggregates to move towards the central hole through water-bearing fractures marked as planes [20].

of about 300 kPa across 1 m grout length can be critical for such soft gels [15].

Grouting methods

Pressure-injection has been successfully used in field tests. However, wall friction can restrict grout penetration depth to a range of about a few decimetres to about a metre for normal constant injection pressures. 'Dynamic' injection using pressure pulses superimposed on a static 'back-pressure' can be used to assist the injection. However, this benefit is limited when injecting clay grouts because thixotropic strengthening of the penetrated grout generates strong flow resistance grout [8]. Electrophoretic injection has been tested with some success but gas production can reduce the effect (Figure 6.45) [20].

6.4   References

1. Pusch, R., Schomburg, J., Kaliampakos, D., Adey, R. and Popov, V., 2004. Low Risk Deposition Technology. Final report European Commission, Contract No EVG1-CT-2000-00020, Brussels.
2. Yong, R.N., 2004. 'On engineered soil landfill barrier-liner systems', Special Lecture, In the *Proceedings of Malaysian Geotechnical Conference 2004*, pp. 109–118.
3. Yong, R.N., 2001. *Geoenvironmental Engineering: Contaminated Soils, Pollutant Fate, and Mitigation*, CRC press, Boca Raton, FL, 307p.
4. Yong, R.N. and Mulligan, C.N., 2004. *Natural Attenuation of Contaminants in Soils*, Lewis Publishers, Boca Raton, FL, 319p.
5. Pusch, R. and Kihl, A., 2004. 'Percolation of clay liners of ash landfills in short and long time perspectives'. *Waste Management and Research*, Vol. 22, No. 2: 71–77.
6. Yong, R.N., Cabral, A. and Weber, L.M., 1991. 'Evaluation of clay compatibility to heavy metal transport and containment, permeability and retention.' In the *Proceedings of the First Canadian Conference on Environmental Geotechnics*, Montreal, pp. 314–321.
7. Pask, J.A. and Davis, B., 1945. Thermal Analysis of Clays and Acid Extraction of Alumina from Clays, Report on Differential thermal analysis, US Bureau of Mines, Denver, Co. 6p.
8. Sposito, G., 1984. *The Surface Chemistry of Soils*. Oxford University press, New York, 234p.
9. Pusch, R., 2002. The Buffer and Backfill Handbook, Part 2: Materials and Techniques. Swedish Nuclear Fuel and Waste Management Co (SKB). Technical Report TR-02-12. SKB, Stockholm.
10. Kenney, T C., Lau, D. and Ofoegbu, G.I., 1984. 'Permeability of compacted granular materials'. *Canadian Geotechnical Journal*, Vol. 21, No. 4: 726–729.
11. Pusch, R., Karnland, O. and Hökmark, H., 1990. GMM – A General Microstructural Model for Qualitative and Quantitative Studies of Smectite Clays. Swedish Nuclear Fuel and Waste Management Co (SKB), Technical Report TR-90-43. SKB, Stockholm.

12. Pusch, R. and Feltham, P., 1980. 'A stochastic model of the creep of soils'. *Géotechnique*, Vol. 30, No. 4: 497–506.

13. Pusch, R., 2000. On the Risk of Liquefaction of Buffer and Backfill. SKB Technical Report TR-00-18, Swedish Nuclear Fuel and Waste Management Co (SKB).SKB, Stockholm.

14. Yong, R.N., Yaacob, W.Z.W., Bentley, S.P., Harris, C. and Tan, B.K., 2001. 'Partitioning of heavy metals on soil samples from column leaching tests.' *Journal of Engineering Geology*, Vol. 60: 307–322.

15. Pusch, R., 1994. *Waste Disposal in Rock, Developments in Geotechnical Engineering*, 75. Elsevier Scientific Publishing Company, Amsterdam, 490 p.

16. Pusch, R., Muurinen, A., Lehikoinen, J., Bors, J. and Eriksen, T., 1999. Microstructural and Chemical Parameters of Bentonite as Determinants of Waste Isolation Efficiency. Final Report EUR 18950 EN, EC Contract No. F14W-CT95-0012, Brussels.

17. Kröhn, K.-P., 2003. 'New conceptual models for the resaturation of bentonite'. *Applied Clay Science*, Vol. 23: 25–33.

18. Gunnarsson, D., Börgesson, L., Hökmark, H., Johannesson, L-E. and Sandén, T., 2002. Installation of the Backfill and Plug Test. In *Clays in Natural and Engineered Barriers for Radioactive Waste Confinement Proceedings of International ANDRA meeting in Reims*, 9–12, December ANDRA, Chatenay-Malabry, France.

19. Börgesson, L. *et al.*, 2004. Plug Tests at the AEspoe URL., Swedish Nuclear Fuel and Waste Management Co (SKB). SKB, Stockholm (Personal communication.)

20. Pusch, R., 1978. Rock sealing with bentonite by means of electrophoresis. Bull. *International Association of Engineering Geology*, Vol. 18, No. 20, Krefeld 187–190.

7 Long-term function of smectite clay for waste isolation

7.1 General

Traditionally, the design of clay-based barriers is the responsibility of geotechnical engineers trained in soil mechanics and its application in foundation and earthquake engineering, slope stability and other issues in ground engineering. Some of the basic tools available and used include: (a) the vast body of classical and engineering mechanics, (b) geomorphology and engineering geology and (c) a range of laboratory tests specifically designed to determine soil properties, characteristics and performance. Except for tests designed to examine properties and performance of soils under extreme weather conditions (high and low temperatures) and other specific constraints, the common practice for laboratory testing has been to perform tests at room temperature under aerobic conditions.

Experience gained from assessment of long-term performance of *in-situ* soil-engineered structures has taught us that: (a) it is not realistic to assume that soil properties determined in the laboratory would remain unchanged in the *in-situ* context, especially when the operative time for the tested soil is in the hundreds and thousands of years and (b) one needs to realize that the soil *in-situ* is in a dynamic environment where abiotic and biotic reactions abound. The performance of soil-engineered barrier–buffer systems, constructed to contain and manage hazardous and non-hazardous wastes faces challenges that fall directly into the long-term performance category. This is particularly evident in the case of chemical leachates interacting with the barrier-backfill materials and with repositories designed to isolate high level nuclear waste (HLW) materials.

Very rarely are chemically induced changes in composition and properties of the soil-engineered barriers considered. To a very large extent, this is because of the lack of understanding of the actual processes involved and suitable means to determine what kinds of mechanisms and processes are involved in the dynamic soil environment. Determination of the rate of change of soil properties and in consequence, the change in expected soil performance is particularly difficult.

A conservative approach is useful in: (a) identifying the most drastic possible changes and (b) determining the corresponding material data required. We will examine two major cases for which altered material properties have the most significant impact on barrier performance, that is, the clay liner portions (clay layers) of the top and bottom barrier–liner systems of waste landfills (Figures 6.1–6.3), and buffer clay embedding canisters with high level radioactive waste (HLW) in underground repositories.

7.2 Clay liners of waste landfills

7.2.1 Clay layer in top barrier-liners

General

We refer to the clay layer in the top barrier-liner of waste landfills (see Figure 6.1). For the purpose of this discussion, and to highlight the performance aspects of the clay layer itself, it is useful to assume the clay layer as the sole liner system. A number of stability problems can occur in top liners that are related to the microstructural constitution of the liners:

- Freezing
- Desiccation
- Erosion
- Chemical impact
- Gas permeation
- Liquefaction
- Change in hydraulic conductivity caused by shear strain related to uneven settlement of the waste
- Slope failure, short- and long-term conditions.

Many of these problems have been mentioned briefly in Chapter 6. We will explore these in further detail in this section.

Frost penetration, freezing

Freezing of porewater in ordinary clay soils takes place at around 0°C or somewhat lower temperature depending on the thermal properties of the clay-water system, and on the chemistry of the porewater. Frost heaving and formation of ice lenses occur in soils if they are frost susceptible and if water for ice lens growth is available. The Casagrande classification of frost susceptible soils considers: (a) well-graded soils where 3% or more of the soil particles are 0.02 mm in particle size and (b) uniformly graded soils where 10% or more of the soil particles are 0.02 mm in particle size, to be highly susceptible to frost heave and ice lens formation – provided that water is available for lens formation and that the appropriate freezing index exists.

The frost heaving pressures created at the ice lens interface with the overlying soil layer can be theoretically calculated. This requires one to

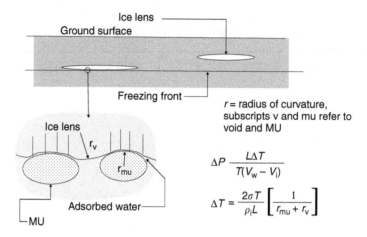

Figure 7.1 Schematic illustration of the local regime at the freezing front showing the curvatures at front surface of an ice lens.

determine the frost heaving pressures as a function of the equivalent radii of representative pore (void) and particle or microstructural unit (MU) – beginning with the Maxwell thermodynamic relations written for a curved ice–water interface as shown in Figure 7.1. From the schematic shown in the figure, since the free energy of the ice must equal that of the water at equilibrium, the reversed curvature of the ice front between pore spaces must be accounted for. Since the minimum potential energy configuration for the ice lens front must be a flat plane, both the ice radii in the pore r_v, and around the MU r_{mu} must be considered. Taking a local region that encompasses an MU and a void space, the relationship obtained [1] shows that the pressure generated between the ice and soil particles or MUs ΔP can be expressed by the Clapeyron relation as:

$$\Delta P = \frac{L\Delta T}{T(V_w - V_i)} \tag{7.1}$$

where

$$\Delta T = \frac{2\sigma T}{\rho_i L}\left[\frac{1}{r_{mu}} + r_v\right] \tag{7.2}$$

and ΔT is the freezing point change at the curved surface, r is the radius of curvature with subscripts v and mu referring to void and MU respectively, L is the latent heat of fusion, V is the volume, with subscripts w and i referring to water and ice respectively, T is the temperature, ρ_i is the density of ice and σ is the ice–water interfacial energy.

From typical electron micrographs, one can deduce the representative radii of curvature for the voids and for the MUs. Acknowledging that there is a distribution of sizes for both void spaces and MUs, one is required to make a judicious choice of a representative unit. Experimental frost heaving tests conducted [1] for fine inorganic silts show that the predictions accord well with measured values of heaving pressures. It is noted that if volume change under frost heaving pressures is allowed, stress relief occurs, resulting thereby in a lower measured heaving pressure. This means that the theoretically calculated values which are based on a no-volume change condition are the conservative maximum values.

A certain amount of local consolidation or compression of the unfrozen material ahead of the freezing front occurs, the extent of which is a function of the water available for ice lens growth and the degree of heave experienced in the layer behind the ice lens. As Figure 7.1 shows, ice lenses will form in different portions and different positions in the soil as the freezing front penetrates into the soil. The positions and extent of ice lens formation is a function of water uptake at the ice lens as it grows, and the rate of freezing [2].

Too rapid a freezing rate will limit the growth of the ice lens, whereas too slow a freezing rate will only create a frozen soil layer, that is, frozen pore-water in the void spaces. The continuity condition that must be satisfied at the freezing front says that the rate of change in thermal energy must be balanced by the spatial rate of change of the heat flux. In this case, the heat flux includes the heat brought in as mass transfer. In all instances, there is an infinitesimal layer of unfrozen water separating the ice lens from the particles and MUs. This water layer is the hydrate layer (or layers, depending upon the activity of the soil particle surfaces) and is often called the adsorbed water layer. Numerous studies have reported on this unfrozen water phenomenon in freezing soils.

With respect to the upper clay soil layer that forms the top liner for a landfill, both freezing and thawing are problems that need to be examined and evaluated. A significant and very evident effect from cyclic freezing and thawing of soils, as in the case of a number of years of exposure of the top clay liner, is the dramatic effect on the reorganization of the microstructure of high clay content soils. Figure 7.2 shows the change in microstructural units from the first freeze–thaw period to the 32nd one. There appears to be a rearrangement of the MUs into more stable units, resulting in the production of wider macropores.

The other primary conclusions reached in the studies on cyclic freeze–thaw effects on soil microstructure and properties of high clay content soils [3, 4] include the following:

- The formation of numerous ice lenses (observed in the test soil columns) in the high clay and water content soils complicates the prediction of depth of frost penetration and frost heaving pressures.

Mineralogy in order
of abundance

Quartz
Illite/mica
Feldspar
Kaolinite
Chlorite
Amphibole

Geotechnical properties

Sp. gr. of solids = 2.68
Dry density = 0.75 g/cm^3
Natural water content = 92%
LL = 69%, PL = 25%

Figure 7.2 SEM pictures showing change in size and shape of MUs in high clay content soil following one cycle (top picture) and 32 cycles of freeze–thaw (bottom picture). Note the scale for both pictures is the same – as shown by the 1 nm scale-bar (black bar) near the centre-bottom for both pictures.

- Repetitive freezing and thawing causes significant variation in the thermal properties of these soils. The production of greater macropores as a result of the cyclical freeze–thaw reorganization of the microstructures seems to be one of the contributing factors.
- The thermal properties of the clays depend substantially on the boundary conditions imposed, particularly on the amount of constraint applied against frost heaving.

Interlamellar water in smectite does not freeze until the temperature is significantly lower than 0°C. Water in external voids (macropores) begins to freeze just below 0°C. This will initiate migration of interlamellar water to the voids and ice lenses as the temperature falls further below 0°C. At very high densities, the amount of interlamellar water is significantly larger than the external water (i.e. water in the macropores) and a large portion of the porewater may remain unfrozen to temperatures below −20°C to −30°C [5]. Table 7.1 shows that smectite resists freezing better than other clay minerals simply because the latter lack interlamellar water.

A practical consequence of the susceptibility to freezing of all clay mineral types is that the clayey component of top liners must be located

Table 7.1 Literature data of unfrozen amount of water
in different soil types

Soil type	−2°C	−10°C
Na montmorillonite-rich clay	40%	35%
Illite-rich clay	21%	19%
Kaolinite	20%	6%
Quartz-rich silt	3%	1%

at frost-free depth. In northern Europe, America and Asia this usually corresponds to a depth of at least 5 m. For very dense smectite clay such as the buffer embedding canisters with HLW, the fraction of interlamellar water is so high that significant freezing will not take place in the foreseeable future conditions – including long periods of permafrost and glaciation.

Desiccation

Desiccation of a clay top liner can cause cracks if the water content drops below the shrinkage limit. It is the breakpoint of the relationship between water content and volume of a drying clay sample, representing the water content below which no further change in volume takes place, that is, when capillary forces are at maximum. Exceeding (going below) the shrinkage limit is associated by a change in colour of the clay. Desiccation cracks can be filled by pervious soil material falling from the overburden, primarily from the adjacent filter that contains silt in its lowest part, and serve as short-circuits of the top liner. If the clay layer is effectively compacted at a low water content and located at about 5 m depth below the ground surface, we should not expect to see continuous desiccation cracks. The low temperature and almost constant moisture regime are expected to discourage desiccation.

Erosion

Physical disturbance by piping and erosion may cause transport of fine particles within and emanating from clay-based top liners. The end result is a more permeable liner. The risk of such degradation, which depends on the flow rate and hence on the hydraulic gradient, has been described in Chapter 5. This risk is highest for liners with low clay contents and negligible for liners with high clay contents. Thus, a liner consisting with about 90% clay fraction such as the Friedland (FIM) clay is not likely to undergo internal microstructural changes. However, the risk of loss of particles downwards requires that it be placed on a bed of silt particles with the void size smaller than about 100 μm.

Geotextiles (GTS) containing a layer of smectite clay that is less than a centimeter thick are particularly sensitive to flow-induced particle transport.

Figure 7.3 shows an example of percolation (permeability) tests with mixtures of 10% smectite-rich clay (MX-80) and 90% ballasts of two types. One is a common ballast material with a steep gradation curve, and the other an improved version obtained by mixing in 20% quartz filler with a grain size ranging between 0.01 mm and 0.1 mm (Figure 7.4). At the end of the tests, the

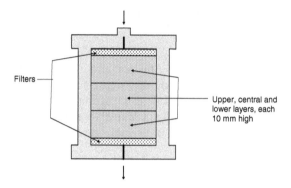

Figure 7.3 Experimental set-up of percolation (permeability) test of 10/90 clay/ballast soil in oedometer.

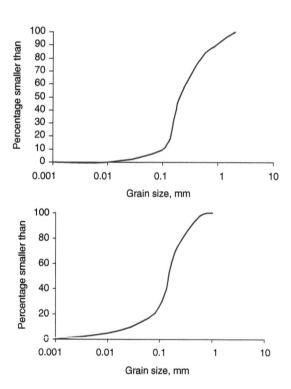

Figure 7.4 Grain size distribution of ballasts A (upper) and B (lower).

soil columns were sliced and the clay content determined by sedimentation tests, the result indicated in Table 7.2. The evident transport of fine particles in the test with Ballast A indicates that the risk of substantial internal erosion is strong if the granulometry of the ballast is unsuitable. The content of finer particles is reduced on the inflow side and increased at the discharge side, meaning that the 'effective' thickness of the liner is decreased. If the soil below the liner has even larger voids, particles will continue to move downward into the voids and may finally exhaust the amount of sealing particles in the liner.

Experimental determination of the hydraulic conductivity in a laboratory is usually undertaken by applying high hydraulic gradients in order to obtain measurable amounts of permeant in a reasonable time frame. A fully water saturated top liner may not be exposed to gradients exceeding 1/10 of those in laboratory tests. However, in the early saturation phase the transient hydraulic gradient may be higher than 10, which may well pose a significant risk of particle transport as illustrated in Figure 7.5.

Table 7.2 Change in content of minus 2 μm particles of a 10/90 mixture in one week percolation test under an average hydraulic gradient of 10 m/m. Dry density = 1600 kg/m³.

Part of sample	Percentage of minus 2 μm particles, ballast A	Percentage of minus 2 μm particles, ballast B
Initially in entire sample	10.0 ± 0.5 (range)	10.0 ± 0.5 (range)
Upper third	7.1 (mean value)	10 ± 1.0 (range)
Central	10.1 (mean value)	10 ± 1.0 (range)
Lower third	11.5 (mean value)	10 ± 1.0 (range)

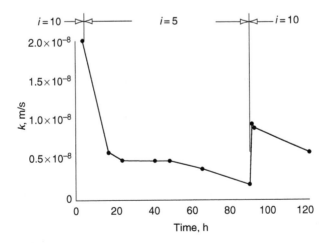

Figure 7.5 Example of the impact of the hydraulic gradient on hydraulic conductivity. Test with mixture of 10% Na-smectite clay and Fuller-graded ballast; dry density = 1200 kg/m³.

Chemical impact

Top liners will normally be saturated with low-electrolyte rain water, and are not expected to undergo significant mineralogical changes for many hundreds or thousands of years. The exception would be if the liner and covering layers contain sulphide minerals and if precipitation is in the form of acid rain. Both would cause a drop in the pH of the porewater. With proper design and construction, the chemical buffering capacity of a thick overburden may be sufficient to maintain a pH value in the range of 6–8.

Gas permeation

Gas transport through a top clay liner has two effects. First, oxygen migrating into the waste mass may generate acid conditions and accelerate release of heavy metal ions from the waste. Gas vents may have to be installed in top liners to release gas at low pressure. They need to serve as back-valves to disallow water entry from above. Second, gas generation induced by chemical reactions in the waste will result in gas overpressure (excess gas pressure), which can lift the entire top liner if the gas conductivity of the clay layer is too low. Penetration of gas may cause permanent channels if the self-healing capacity (expandability) of the clay is insufficient. Several models for calculating gas migration through expansive clays have been proposed – such as the one suggested by Horseman and Harrington [6] where the processes modelled account for the existence of

- a threshold effect that requires the gas pressure to rise above a critical value before migration can begin;
- a propagation phase during which the gas front moves along capillary-like paths;
- pressure release and relaxation phases at and after breakthrough.

Conceptual and theoretical modelling imply that the gas permeation starts in independent gel-filled channels distributed randomly at the inlet face of the clay when the gas pressure is in the same order of magnitude as the bulk swelling pressure. Whereas the original primary channels are directed parallel to the flow direction, subsequent generated channels have different directions. This is a realistic feature for relevant modelling of gas and water migration. Under steady state conditions, flow is expressed by the Poiseuille law. The flow rate is determined by the geometry and gas viscosity. This model gives results that are in reasonable agreement with experimental data [7, 8]. For practical purposes one can assume – although this still has to be proved – that a smectite content of about 50% is sufficient to produce self-sealing even after long-term peristaltic gas penetration. The most important issue in the design of top liners is to ensure that gas can be released through the liner without lifting it, meaning that the critical gas pressure must be lower than the overburden pressure.

Liquefaction

Liquefaction may be a serious problem for any low density water-saturated landfill. Seismically induced shearing will cause contraction and development of a porewater overpressure (excess porewater pressures) that can reduce the effective pressure to a critically low level, particularly in slopes. For densities exceeding about 1800 kg/m³, this is not expected to be a problem as long as the expected seismic events represent values lower than 6 on the Richter scale [9].

7.2.2 Bottom liners

Bottom clay-based liners will also be exposed to hydraulic gradients that can cause migration of clay particles. Their role in the total engineered barrier system can be seen in Figures 6.1–6.3. As stated in Chapter 6, for the purpose of evaluating clay liner performance, we assume that the liners mentioned earlier have been breached and that leachate penetration into the clay liner will occur. The gradients in the bottom liners can be higher than for top liners and the risk for internal transport of fine particles consequently higher. However, the impact of contaminants in the porewater released from the waste is more important, and as indicated in Chapter 6, compatibility between liner and the chemistry of the leachate needs careful consideration in bottom-liner design. The basis for this has been outlined in Chapters 2 and 3, and developed in further detail in Chapter 6. At this stage, we will provide a few practical examples of the performance of layers representing bottom liners tested in the laboratory under hydraulic and temperature gradients.

The hydration phase

Table 7.3 shows that the electrolyte content of the water that moves in from the hydraulic boundaries has a significant impact on the saturation rate. It shows the outcome of 45-day laboratory tests where wetting with salt

Table 7.3 Water saturation of MX-80 clay samples exposed to a temperature gradient of 15°C/cm and a water pressure of 50 kPa at the cold end for 45 days [8]

Distance from cold end (mm)	Water content after 45 days for 3.5% NaCl solution	Water content after 45 days for 3.5% CaCl₂ solution
10	47	45
20	45	41
30	32	38
40	12	33
50	8	28

solutions was allowed from one end of 50 mm long cells with MX-80 clay. The temperature at the wet end was 20°C and 90°C at the opposite end. The clay had an initial water content of 10% and a dry density of 1270 kg/m³. Experiments with distilled water showed noticeable wetting to only 30 mm distance in this 45 day time period – in direct contrast to the tests with NaCl and CaCl₂ solutions where complete penetration and egress of the solutions were obtained. This shows that the microstructure of the hydrating clay is significantly affected by the solutes.

The effects are explained by colloid chemistry: Ca causes contraction of the stacks of lamellae yielding larger pore spaces. The diffuse double-layer (DDL) model shows that aggregation of particles will occur in salt water. The two effects combine to give a higher hydraulic conductivity – as compared to the soil wetted by electrolyte-poor water.

Percolation (permeation)

Permeability tests on mixtures of clay/ballast soil conducted with a high pH solution show some very interesting results. Two mixtures of the ballast termed B (cf. Figure 7.4) were prepared with 5% and 10% MX-80 clay, the dry density being about 1950 kg/m³ for both. The Ca-rich solution with a total content of dissolved salts of 19.5% and pH 12.4, originated from the preparation of a slurry of ash and cement to be pumped out in a basin. The conductivity was determined after: (a) initial saturation of the mixtures with distilled water (b) permeated with distilled water for 3 days and (c) subsequently permeated with the high-pH water (Figure 7.6).

The test with 10% MX-80 clay showed a typical low conductivity performance when permeated with distilled water. However, this changed

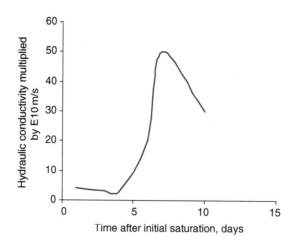

Figure 7.6 Change in hydraulic conductivity of 10/90 smectite/ballast mixture with B-type ballast (cf. Figure 7.5).

when a high pH solution was introduced at day 4 as a replacement for the distilled water. The increase in conductivity from day 4 when the salt solution began to replace the non-salt porewater, that is, from $K = 2E{-}10$ m/s to $K = 5E{-}9$ m/s on day 7, was the result of exchange of Na by Ca. The two effects of note are: (a) decreased particle interactions in accord with changes in DDL forces and (b) possible aggregation of particles. From day 7 onward, the decrease in hydraulic conductivity suggests that dissolution of the smectite is taking place, and that the dissolution products such as amorphous silica/aluminium complexes are responsible for void plugging.

The mixture with 5% MX-80 clay gave similar trends. In the first phase, when distilled water was used as the permeant after saturation, the conductivity was about $6E{-}9$ m/s. It then rose to $1.1E{-}8$ m/s, indicating a dramatic increase that is believed to be due to decreased DDL interactions and strong aggregation of particles resulting in a near total loss of sealing capacity of the clay component. The small amount of clay in this test which was vulnerable to the reaction effects from high the pH may have converted to non-swelling reaction products in a few days.

It is important to realize that true mineralogical changes in the form of conversion of smectite minerals to non-expandable forms, as well as dissolution and precipitation of secondary clay minerals and calcite, can occur at ambient temperature in bottom liners. The influence of increased temperature that can develop in certain waste types will serve to accelerate the reactions. The very significant importance of temperature for mineral conversion makes it important to consider it in detail.

7.3 HLW isolation – a type of temperature-related mineral conversion

7.3.1 *General*

Hydrothermal conditions can cause conversion of smectite minerals to illite and chlorite. The best example of this is the well-documented successive alteration of deeply located sediments in the Gulf area. This change is associated with changes in the void size distribution and release of silica and aluminium. The silica easily migrates and precipitates in regions where the temperature is lower, while the aluminium is more or less retained where it was released. These changes are expected to occur also in bottom liners if the waste produces heat, and in the buffer surrounding hot canisters in HLW repositories.

7.3.2 *Reactions under hydrothermal conditions*

Illitization

Various investigations have shown that conversion of smectites to non-expandable clay minerals, primarily as illite (hydrous mica) and chlorite.

The exact process is not known but two reactions have been proposed: (1) successive intercalation of illite (I) lamellae in smectite (S) stacks forming mixed-layer (IS) minerals [10, 11], and (2) neoformation of illite by co-ordination of Si, Al, K and hydroxyls to an extent controlled by access to either of them. The general reaction is assumed to be:

$$\text{Smectite} + K^+ + Al^{3+} \rightarrow \text{Illite} + Si^{4+} \tag{7.3}$$

This type of process is believed to be valid for the entire smectite family but the activation energy for conversion and dissolution is different and for beidellite it may be that potassium uptake can directly lead to collapse of the stacks of lamellae to yield illitic minerals. Although montmorillonite is the most common smectite species in nature, and has the best gelation properties, extensive experience in deep drilling shows that saponite is not affected by calcium and salinity as much as montmorillonite, and is considered to have a higher chemical stability at high temperature. This points to a higher activation energy for conversion of saponite to illite in comparison to montmorillonite.

The numerous examples of illitization provided by the nature do not call for further documentation. One of the best and most discussed reference cases is the Kinnekulle 'metabentonite', so-called because of the significant illitization of the montmorillonite-rich clay originally formed from volcanic ash [12]. A series of Ordovician bentonite layers with a main bed of 2 m thickness is located about 100 m below a sub-horizontal diabase cap rock that heated the entire sediment series for several hundred years. The bentonite layers are concluded to have been exposed to a temperature of up to 160–185°C. Detailed studies have been conducted on the clay mineral content and on physical properties such as permeability, shear strength and creep behaviour. The uppermost part of the 2 m thick bed and adjacent thinner layers are significantly silicified, and the clay fraction consists of I/S minerals with 30–60% smectite. The central part contains 50–70% smectite in the I/S clay and a separate illite phase. This fits well with theoretical calculations of the reaction rate presently applied in Swedish R&D work for HLW isolation based on Pytte's theory [13]. It is a thermodynamically based Arrhenius-type theory that assumes that the rate of change of the smectite-to-illite ratio depends on temperature, potassium content in the porewater and activation energy. The activation energy is commonly taken to be 25–30 kcal/mole and for the Kinnekulle case, 27 kcal/mole would fit the theory. The criterion of sufficient access to potassium can be fulfilled by assuming that potassium diffused from groundwater in the sediment series. The assumption of strong groundwater convection is supported by the existent temperature conditions. The rate equation is:

$$-dS/dt = \left[Ae^{-U/RT(t)}\right]\left[(K^+/Na^+)mS^n\right] \tag{7.4}$$

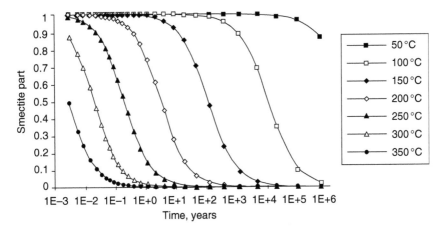

Figure 7.7 Diagram showing the expected conversion of smectite to illite for the activation energy 24 kcal/mole according to Pytte's theory. 'Smectite part 1' represents 100% smectite, which drops to 97% in E5 years at 50°C and to 50% in 1000 years at 200°C [8].

where S is the mole fraction of smectite in I/S assemblages, R is the universal gas constant, T is the absolute temperature, t is the time and m, n is the coefficients.

Figure 7.7 shows the rate of illitization for an activation energy of 24 kcal/mole to yield conversion from smectite. For this value, the model implies that heating to 50°C causes only an insignificant loss of smectite in E6 years, while about 100% of the original smectite turns into illite in this period of time at 100°C. Buffer clay in the HLW repository is exposed to heat for a period lasting at least a few thousand years, with a typical temperature history being: (a) 100–150°C in the first hundred years, (b) followed by a 500-year period with an average temperature of 50–100°C and (c) less than 50°C in the subsequent 1000 years. For this scenario, about 85% of the smectite would remain after about 1500 years and more. Comprehensive work has been undertaken to determine if Pytte's model adequately describes S/I conversion – as may have occurred historically in nature.

Silica precipitation

Conversion of smectite minerals to non-expanding ones, primarily illite, is associated with release and migration of silica according to Eq. (7.3) and precipitation of silica when lower temperatures prevail, resulting in cementation effects. The detailed process in the release and precipitation of silica is not known but it may be related to conversion of montmorillonite in 'low-temperature' form, that is, the Edelmann and Favejee crystal structure, to the 'high-temperature' form represented by the Hofmann, Endell and

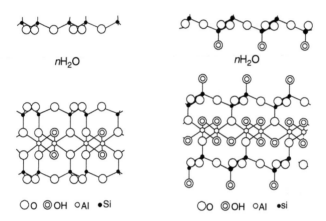

nH_2O nH_2O

Oo ⊚OH oAl •Si   Oo ⊚OH oAl •si

Figure 7.8 Possible lattice constitutions of montmorillonite. Left: Hofmann, Endell and Wilm version ('high-temperature' form). Right: Edelmann and Favejee version ('low-temperature' form) [14].

Wilm structure (Figure 7.8) [14]. This may imply replacement of silicons from the tetrahedral layers by aluminium ions originating from cannibalized smectite minerals, or accessory minerals. Silica would appear in ionic form as H_4SiO_4 and precipitate as amorphous silica, cristobalite or quartz. The process has been identified in nature in all cases where diabase dikes have intersected bentonite beds. The Kinnekulle case is a good example. Recently, a detailed analysis was made of a Libyan case where a 5 m wide Tertiary diabase dike intersected up to 5 m thick sub-horizontal bentonite layers, also of Tertiary age [15].

The temperature that the bentonite has been exposed can be estimated by optical and scanning electron microscopy (SEM), using Lindsley's pyroxene thermometry. Application of this technique to the Lybian case produced results that gave a maximum temperature of the bentonite next to the diabase of 500°C after 1 month, and of 200°C after 1 year. At about 3 m distance the maximum temperature had been less than 100°C.

The following major conclusions to be drawn are:

- The cooling rate was very high.
- A silicified zone was found next to the diabase contact and fluid inclusions in it contained hypersaline solutions.
- The saline fluid circulation led to transformation of a small part of the montmorillonite to kaolinite near the diabase.
- CEC dropped with increasing distance from the diabase (Figure 7.9).
- Fractures indicating brittleness of the clay within 0.15 m from the diabase, the brittleness being caused by silicification, and the fracturing by thermo-mechanical stresses.

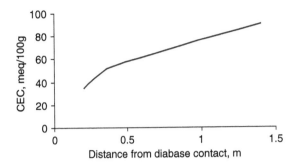

Figure 7.9 CEC of the Libyan montmorillonite bentonite at different distances from the diabase contact [15]. The curve asymptotically approaches the value 100.

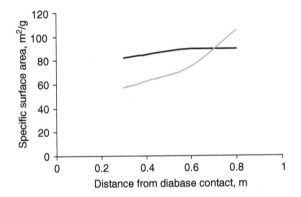

Figure 7.10 SSA of the Libyan bentonite. Thin curve: S_{BET} representing basal surfaces of stacks of lamellae. Fat curve: S_{micro} representing the interlamellar space [15].

- At about 3 m distance from the diabase contact, where the temperature had not reached 100°C, no mineral changes were found.
- The specific surface area (SSA) representing the interlamellar space had dropped close to the diabase, while the surface area representing the basal surfaces of the stacks of lamellae was not changed.
- The grain size distribution indicated coarse, cemented particle agglomerates near the diabase. Thus, at 4 m distance the minus 2 μm content was 65% whereas it was 30% at 0.2 m distance (Figure 7.10).

Comprehensive microstructural studies using SEM and TEM have been performed to identify and characterize silica precipitations formed in hydrothermally treated montmorillonite-rich clays [16]. These studies were made by confining samples that had been saturated with distilled water and homogenized in gold-plated autoclave cells, and kept in ovens at temperatures ranging from room conditions to 200°C for a few weeks to one year. Examination and analyses of presumable precipitates by EDX indicated that they were cristobalite quartz of poor crystallinity, and amorphous silica, while in nature, as illustrated by the very slowly cooled Kinnekulle bentonite, the precipitates were well crystallized quartz. Figures 7.11, 7.12 and 7.13 illustrate the change in microstructure of dense smectite clays by hydrothermal treatment, that is, permanent collapse of the stacks of lamellae, yielding dense stacks packets separated by larger voids than in untreated clay.

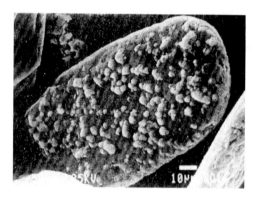

Figure 7.11 Silt particle coated with small quartz particles believed to have been formed by precipitation of silica emanating from smectites in Kinnekulle bentonite that was heated to an average temperature of 160°C for a few hundred years [17].

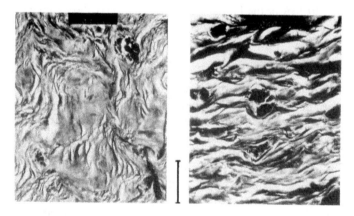

Figure 7.12 Comparison of montmorillonite-rich clay at room temperature (left) and after hydrothermal treatment at 150°C for 1 year (right). Notice the collapse of stacks of lamellae and formation of dense siliceous particles in the heated clay. Dry density = 1590 kg/m³. Bar is 1 μm [16].

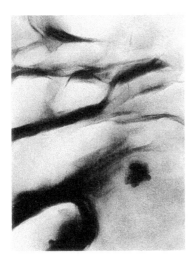

Figure 7.13 Microstructural changes in the form of collapse of stacks of lamellae and precipitation of dense siliceous particles at 200°C treatment. Dry density = 1590 kg/m³ [16].

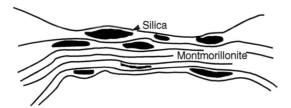

Figure 7.14 Distribution of precipitated silica [18].

Precipitation of silica within particle aggregates is believed to weld them together. This will increase their strength and stiffness whilst reducing ductility and enhancing brittleness (Figure 7.14).

Practical importance of silicification

The impact of silicification has been clearly demonstrated through uniaxial compressive testing of hydrothermally treated montmorillonite (Figure 7.15). The axial load on the samples, which were shielded to avoid desiccation, was increased stepwise, leaving each load on for 5 minutes to allow creep strain to develop. The dry density was 1590 kg/m³, that is, 2000 kg/m³ in the actual water saturated form. The illustration shows that the untreated reference clay sample failed when the uniaxial stress was raised from 1.30 to 1.74 MPa, while the samples that had been heated to 105°C and higher temperatures had become stronger and stiffer. The one heated to

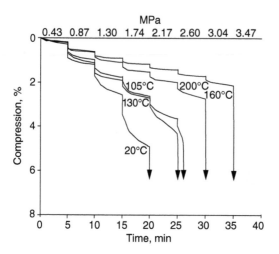

Figure 7.15 Results of uniaxial testing of hydrothermally treated montmorillonite clay [8].

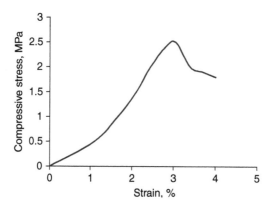

Figure 7.16 Example of uniaxial compression testing. The density of the saturated clay was about 2100 kg/m³.

180°C had a strength that was more than 2 times that of the reference sample, and it also showed a correspondingly lower compressive strain. Since cementation, increased activity by the DDL forces, and microstructural reorganization all combine to cause strengthening, it is not possible to determine which of the processes is most important.

The stress/strain behaviour of a totally illiticized and silica-cemented bentonite clay from Burgsvik, Sweden, is used to illustrate the mechanical properties of the end product of buffer clay (Figure 7.16). A study of the thermal, hydraulic and mechanic (THM) history of the Ordovician clay indicates that it was exposed to a temperature of 50–100°C for several million years. According to Figure 7.7, one would expect that this would have been completely illiticized.

The results indicate that failure developed at an axial strain of about 3% for a compressive stress of 2.5 MPa. The peak suggests cementation and brittleness of the claystone sample – with some ductility – as evidenced by the large strain. The obvious similarity of the behaviour of the illitized Burgsvik clay and that of the hydrothermally treated but mineralogically unchanged clay shown in Figure 7.15, demonstrates that the strength increase and brittleness of the latter were caused by silicification and microstructural reorganization and not by illitization.

7.3.3   Chemical mechanisms in illitization and silica precipitation

Basic processes

Several attempts have been made to model chemical alteration of mont-morillonite under repository-like conditions. In this section, we will describe and apply a recent model proposed by Grindrod and Takase [19] to assess the practical importance of the derived changes in clay proper-ties. The model focuses on dissolution of smectite and precipitation of illite, and is in better agreement with the actual smectite/illite (S/I) conversion and precipitation of silica of the Kinnekulle bentonite bed than solid–solution models [18]. In contrast to the conventional Pytte (Reynold)-type models, the one proposed by Pusch *et al.* [20], takes $O_{10}(OH)_2$ as a basic unit and defines the general formula for smectite and illite as:

$$X_{0.35}Mg_{0.33}Al_{1.65}Si_4O_{10}(OH)_2 \quad \text{and}$$
$$K_{0.5-0.75}Al_{2.5-2.75}Si_{3.25-3.5}O_{10}(OH)_2 \tag{7.5}$$

where X is the interlamellar absorbed cation (Na) for Na-montmorillonite. The reactions in the illitization process are:

Dissolution

Dissolution takes place according to Eq. (7.6):

$$Na_{0.33}Mg_{0.33}Al_{1.67}Si_4O_{10}(OH)_2 + 6H^+$$
$$= 0.33Na + 0.33Mg^{2+} + 1.67Al^{3+}4SiO_{2(aq)} + 4H_2O \tag{7.6}$$

Precipitation of illite and siliceous compounds

Precipitation takes place according to Eq. (7.5):

$$KAlSi_3O_{10}(OH)_2 = K^+ + 3Al_3^+ + 3SiO_{2(aq)} + 6H_2,$$
$$\text{(log } K \text{ being a function of temperature)} \tag{7.7}$$

The rate of the reaction R is taken as:

$$R = A \exp(-E_a/RT)(K^+)S^2 \tag{7.8}$$

where A is the coefficient, E_a is the activation energy for S/I conversion, R is the universal gas constant, T is the absolute temperature, K^+ is the potassium concentration in the porewater and, S is the SSA for reaction.

In a smectite clay buffer embedding a canister with HLW, the thermal gradient will cause more dissolution in the hot clay at the canister surface than at the clay/rock interface. Transport of dissolved species will be towards the cold side.

The chemical code

The chemical model described in the previous section is now coupled with the transport problem in the one-dimensional cylindrical coordinate representing the buffer material. The kinetic reactions are linked with diffusion-dominated transport of the aqueous species, that is, silica, aluminium, sodium, magnesium, and potassium to form a set of quasi-nonlinear partial differential equations for the aqueous species and minerals which fall in the following general class:

$$\frac{\partial c_i}{\partial t} = \nabla D_i \nabla c_i - \sum_j \lambda_{ij} R_{ij}$$

$$\frac{\partial m_k}{\partial t} = \sum_j R_{jk} \tag{7.9}$$

where c_i is the total concentration of aqueous species including element i, m_k is the abundance of mineral k, R_{ik} is the precipitation of mineral k that includes element i and λ_{ik} is the stoichiometry of element i in mineral k.

One then spatially discretizes Eq. (7.9) by a finite difference scheme to obtain a set of ordinary differential equations for each numerical grid. All the aqueous speciation reactions are assumed to be in local instantaneous equilibrium. Based on this assumption, one can determine the concentration of each aqueous species by solving the mass action equations with the total concentration of the element that can be specified by solving Eq. (7.9). A system of differential and algebraic equations (DAE) is obtained that is potentially stiff, that is, including multiple processes of hugely variable time scales, and solved numerically by applying the backward difference formulae to guarantee accuracy and efficiency at the same time [21]. Temperature dependence of the equilibrium constants and the rate constants for the mineral reactions and the aqueous speciation is defined in the same manner.

7.3.4   Application of the model to the KBS-3V case

The hydration phase

There are two effects of a rise in temperature: (1) the viscosity of water entering the buffer clay is reduced (2) the hydration potential is affected. The first effect speeds up flow and diffusion, and the second causes an increase in the average d-spacing (interlamellar spacing) of smectite when the temperature is raised from 20°C to 90°C. This means that the sorption energy drops. Hence, the driving force for hydration of the buffer decreases with increasing distances from the rock.

Basic data

The KBS-3V deposition hole is used as an example in modelling with the following assumed conditions:

- Dry density = 1700 kg/m^3
- Porosity = 0.35
- Montmorillonite fraction is 75%
- Initial pore water is seawater
- Groundwater is of seawater origin and assumed to be in equilibrium with granite at the specified temperature in the surrounding rock, for example 60°C during the first 500 years
- Temperature at canister surface = 90°C, temperature at rock wall = 60°C
- Diameter of hole is 1.8 m, diameter of canister is 1.0 m
- Vertical water flow rate in the 1 cm rock annulus (EDZ) around the hole is 1 mm/day and 1 mm/month, respectively
- Diffusion of K$^+$ in the clay is E–9 m^2/s
- Considered time = 500 years.

Under the prevailing thermal gradient, which is conservatively taken to be constant during the considered time period, silica is released and transported from the hottest to the coldest part of the clay buffer.

Results of calculation

Despite the conservative assumption of constant thermal gradient in the first 500 years, no illite is formed. Furthermore, one finds that no precipitation of quartz or amorphous silica occurs in the clay since both minerals are under-saturated (Figure 7.17). However, there will be a significant loss of quartz due to dissolution/transport in the inner hot part, as illustrated in Figure 7.18.

The results indicate that the presence of the temperature gradient alone is not sufficient for precipitation of quartz or amorphous silica. However, cooling will create a different result. To illustrate this, additional calculations

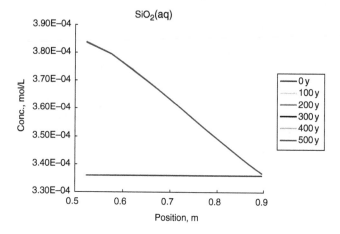

Figure 7.17 Evolution of SiO$_2$(aq) concentration profile (steep curve ~500 years).

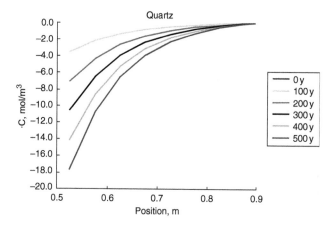

Figure 7.18 Evolution of quartz abundance profile (lowest curve ~500 years).

can be made assuming the same temperature conditions for 500 years, followed by a linear temperature drop with time to 25°C after 10 000 years. The results from this new set of calculations are shown in Figures 7.19 and 7.20. These indicate that precipitation of quartz will occur within about 0.1 m from the rock wall, whereas amorphous silica will not be precipitated. The possibility of amorphous silica precipitation appears to be very much dependent on the time scale of cooling relative to that of quartz precipitation. Rapid cooling leaves the system super-saturated with quartz. The excess concentration of silica in solution eventually reaches the solubility of amorphous silica at the lower temperature. This might be the case in laboratory experiments where cooling is much more rapid than expected in a repository environment.

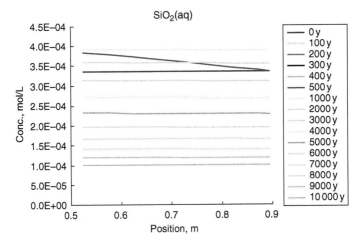

Figure 7.19 Evolution of SiO$_2$(aq) concentration profile (lowest curve ~10 000 years).

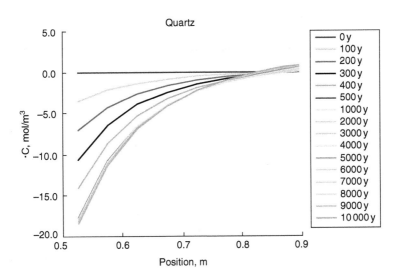

Figure 7.20 Evolution of quartz abundance profile (lowest curve ~10 000 years). Precipitation leading to cementation takes place in the outer and colder 0.1 m part of the buffer.

Implications of the results

General In the example used for discussion, it is clear that cementation by precipitation of quartz will not take place in a 500 year time frame in the KBS-3V buffer clay under the assumed temperature gradient. However, subsequent cooling to ambient temperatures would cause formation of quartz. In a 10 000 year time frame, extensive quartz precipitation would be expected in the larger part of the KBS-3V buffer clay. The impact it may have on the physical properties of the buffer clay has been discussed previously in this chapter, and will be demonstrated by the experimental investigations described in this section.

Application of the Takase/Benbow model for hydrothermal conversion of smectite-rich clay shows that practically no illitization will take place in the first 500 years in the KBS-3V buffer under the previously assumed temperature and groundwater chemical conditions. An increase in solid silicious compounds, primarily quartz, will occur in a large part of the buffer clay in a 10 000 year time frame. This assumes that the temperature remains constant at 90°C at the canister surface and 60°C at the rock wall for the first 500 years, followed by a linear drop of temperature from this state to room temperature in 10 000 years. The expected impact of the physical properties of the KBS-3V clay buffer in the considered 500-year period can be summarized as follows:

- Practically no change in hydraulic conductivity since illitization is negligible.
- Increase in shear strength and deformation moduli by around 50% caused primarily by permanent microstructural reorganization, meaning that the creep rate of the heavy canister embedded in the clay will drop, and that the self-sealing potential drops.

In a 10 000 year perspective, the major changes would be:

- Drop of hydraulic conductivity by precipitation of silica in the voids.
- Dissolution of the smectite buffer manifested by the significant loss of silica.

This provides the elements required for neoformation of illite, which can be extensive depending on the conditions for migration of potassium to the buffer.

Practical importance of illitization Conversion of smectite to non-expanding minerals means that part of the porewater that is originally strongly sorbed in the interlamellar space, will be released into widening voids. For a constant bulk density, this can increase the hydraulic conductivity by a few orders of magnitude, and greatly reduce the expandability and self-sealing potential of the material.

Practical importance of cementation The impact of silica precipitation on the hydraulic conductivity of montmorillonite-rich clay buffer with a density intended for the KBS-3V concept is not believed to be very significant but brittleness and loss in self-sealing ability will be obvious. The possible increase in conductivity in the hottest part is probably much less important than that caused by microstructural rearrangement, that is, widening of voids between the stacks due to collapse of the stacks of lamellae. Experience from previous laboratory tests of samples extracted from the MX-80 clay in the Stripa Buffer Mass Test suggest that the changes in the hydraulic conductivity would be minor in the buffer clay used in the chosen example (KBS-3V). The test temperature of 125°C at the canister (heater) surface and 72°C at the rock wall in the Stripa Buffer Mass test series were maintained for the test period of 0.9 years. For this specific set of conditions, and in respect to the influence of silica precipitation and cementation, the conductivity and swelling pressure of the samples could not be distinguished from those of unheated MX-80 [22].

7.3.5 Impact of salt water on the crystal structure

Conversion of smectite to illite has been the focus of numerous scientific investigations. The most recent study of interest has shown evidence in natural geological systems of low-temperature illitization of smectite in contact with saline solutions, for example, in the Slowak basin. The reaction, however, is very slow at low temperatures. Higher temperatures will change this. The 30-day laboratory studies reported by Kolarikova and Kasbohm *et al.* [15, 23] conducted using MX-80 bentonite provide us with some useful information. The MX-80 bentonite at 1300 kg/m^3 dry density was saturated with 10% and 20% NaCl solutions to detect possible signs of early mineral alterations. The experiments were conducted at 110°C to accelerate reactions. At the termination of the tests, analyses showed that the main component was still montmorillonite, with largely preserved expandability when determined with ethylene–glycol treatment (16.9 Å). However, Al-enrichment in the octahedral layers had occurred as shown by TEM-EDX examination, and substitution of Si by Al in tetrahedral layers had taken place as well. Beidellite had been formed and a small 10 Å-interference of XRD spectra indicated possible formation of illite/brammalite.

7.3.6 Impact of salt water and temperature on rehydration

The rehydration potential of montmorillonite after drying in air changes with temperature, as illustrated by tests on Ca/Mg-montmorillonite. Figure 7.21 shows that this effect is obvious for clay heated to more than about 100°C. This raises the question of the significance of the safety margin for a buffer heated to 90°C [15]. Loss of the least strongly held interlamellar water is believed to occur at 70°C whereas total loss of the remaining interlamellar water occurs at about 130°C. The temperature 100°C can be assumed to represent a critical stage where permanent collapse of stacks with incomplete occupation of the interlamellar space results.

Cementation by precipitation of silica released from the lattice tetrahedrons may contribute to irreversible contraction. This is possible as a conversion from the 'low-temperature' Edelmann/Favejee type to the 'high-temperature'

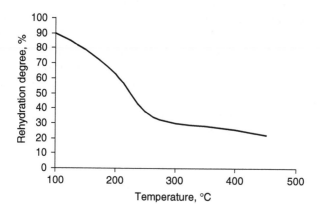

Figure 7.21 Rehydration potential of Ca/Mg montmorillonite [15].

Hofmann/Endell/Wilm structure – a process representing different atom co-ordination stages.

7.3.7 Interaction of cement and smectite clay

Chemical reactions

Deep geological disposal of waste will utilize large amounts of concrete because drifts, tunnels and shafts in the host rock require support in locations where the rock is unstable. The use of ordinary Portland cement produces gives a very high pH product. This high pH will degrade smectite clay in contact with the concrete product. Theoretical considerations [24] and studies using batch tests with $KOH/NaOH/Ca(OH)_2$ at 90°C have indicated that the reaction products would be zeolites, such as phillipsite and analcime, and that they would be formed in a few months [25]. Interstratified smectite/illite (S/I) with up to 15–20% illite was found when the solution contained much K^+. Uptake of Mg^{2+} in the montmorillonite crystal lattice resulted in the smectite species saponite. The most important conclusions from these studies indicate:

- High-alkali cement degrades quicker than low-alkali cement. The latter deteriorates by destruction of the CAH gel.
- Dissolved elements and water migrate from the fresh cement paste to contacting smectite in the first few hours. Ca migrates from the cement to the clay causing ion exchange and change in the microstructure of the clay by coagulating softer parts.
- The cement paste is dehydrated and its voids become wider. Water moves from the smectite to the dense cement matrix.

- The dense cement paste fissures.
- pH = 12.6 is a critical value for significant changes of the smectite component in a short time frame. For longer periods of time, a lower pH is believed to cause permanent mineral alterations.

Because of the fear that Portland cement in a concrete bulkhead can produce a large high-pH plume that can affect smectitic backfills at a distance of several meters, various research projects have addressed the development of low-pH cements for use in repositories containing smectite clay. The results from two recently developed cement types interacting with FIM clay [26] provide the basis for the discussion in this section.

Test performance

The experiments were conducted using the types of cells shown in Figure 7.22. Clay powder was compacted in the cells to a density corresponding to about 1900 kg/m^3 after fluid saturation. The central perforated tube was surrounded by a thin filter to prevent clay particles from migrating into the cement–water solution that was contained in the tube. The solution was replaced at the end of each test period (two periods of 2 months duration followed by a 1 month long period) and analysed with respect to the concentration of pH and important elements. After 5 months, parts of the clay samples were extracted and examined using X-ray diffraction (XRD) and electron microscopy. The rest of the clay samples were used for determination of the hydraulic conductivity after filling the tube with presaturated FIM clay.

Clay

FIM clay, known by geologists as Friedland Ton, contains the following major elements expressed in oxide form and weight percent: $SiO_2 = 57.2$,

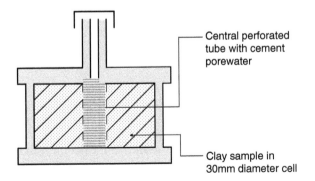

Figure 7.22 Cell for clay/cement–water experiments.

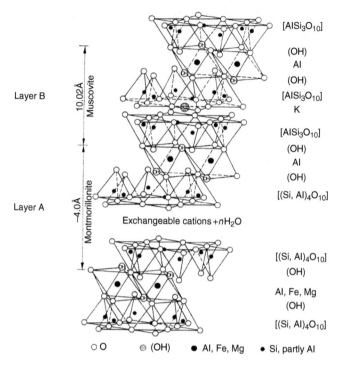

Figure 7.23 Crystal structure of Friedland Ton muscovite-montmorillonite. It can be described as a system of alternating layers of irregular sequences (e.g. ···AABAAABBABAAABAA···). The proportion between A and B is 70% A (montmorillonite) to 30% B (muscovite). The CEC is about 50 meq/100 g. Organic molecules can be contained and verified by XRD analysis.

$Al_2O_3 = 18.0$, $Fe_2O_3 = 5.5$, $MgO = 2.0$, $K_2O = 3.1$, $Na_2O = 0.9$, $S = 0.3$. The adsorbed cations in meq/100 g are $Na^+ = 11.0$, $Ca^{2+} = 4.0$, $Mg^{2+} = 2.0$, $Fe = 0.49$, $CEC = 50.0$ meq/100 g. Figure 7.23 shows the crystal structure of this mixed-layer mineral.

Cement

Two 'low'-pH cements were used with the chemical composition expressed in oxide form of the major components shown in Table 7.4. The slag cement MERIT 5000 is produced by the Swedish Merox (SSAB) company and is commonly used for ordinary construction work. The calcium aluminate ELECTROLAND cement is manufactured by the Spanish company Cementos Molins Industrial SA using limestone and bauxite as raw materials. It is also used for construction purposes and is reported to be slow-hardening and workable up to 4 hours after preparation.

Table 7.4 Chemical composition of the Si-rich MERIT 5000 and the Al-rich ELECTROLAND cements

Type	SiO_2	Al_2O_3	MgO	Na_2O	CaO	MgO	K_2O	Fe_2O_3 FeO
M[a]	34	13	17	0.5	31	17	0.5	0.3
E[b]	3	41	—	—	39	—	—	16

Notes
a M = MERIT 5000.
b E = ELECTROLAND.

Table 7.5 Cement porewater composition in mg/l in the 5-month test

Type	Time, months	Si	Al	Mg	Na	Ca	K	Fe	Ph
M[a]	0	3.2	0.3	16.9	1260	30	107	0.2	9.4
	2	5.8	0.9	51.0	2030	155	82	1.8	7.8
	4	5.0	0.6	31	1450	69	99	1.0	7.9
	5	6.2	0.6	22.7	1070	74	112	1.0	7.9
E[b]	0	1.9	2.5	19.2	1540	39	117	0.3	8.1
	2	3.7	20.7	21.8	1890	113	96	6.1	7.9
	4	3.5	35.7	27.9	1450	78	112	1.0	8.0
	5	6.7	19.1	27.0	1130	219	98	5.3	8.1

Notes
a M = MERIT 5000.
b E = ELECTROLAND.

Chemical interaction

The reactions identified were as follows:

- Slight dissolution of the clay was manifested by increased Si, Al, Mg and Fe contents in the solutions.
- An increase in iron in the solutions indicated that dissolution of the chlorite component of the clay appears to have dominated the slight degradation of the clay (Table 7.5).
- An increase in Ca and drop in Na concentration indicated cation exchange by uptake of Na from the solutions by the clay and release of Ca to them.
- The insignificant change in K-content in the solutions suggests that illitization was absent or negligible.
- XRD (Figure 7.24) analyses verified that no significant changes in the mineralogical composition were observable and that zeolites had not been formed although X-ray amorphous particles of such minerals may have been present.
- Electron microscopy showed the typical structure of Friedland Ton, with no minerals of zeolite-type present. The element distribution spectra showed a Ca-peak only from the interaction with Merit 5000.

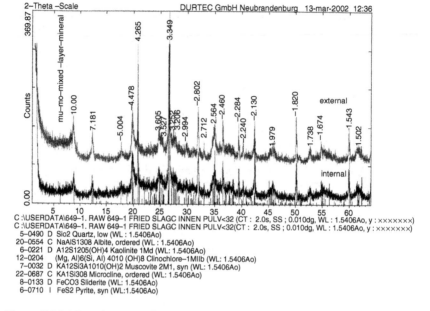

Figure 7.24 XRD diagram of FIM clay powders from outer ('external') and inner ('internal') parts reacted with porewater of Merit 5000.

- Saturation and exposure to cement water in the 5-month long test period did not markedly change the hydraulic conductivity except that the treated clay had its hydraulic conductivity reduced to about 1/4 of the value of untreated clay. This reduction can have been caused by clogging of voids by amorphous hydrosilicate compounds formed by cement–water attack on the clay minerals, or by flow-generated transport of clay aggregates to clogging positions in microstructural voids. A safe conclusion is that no significant change in bulk hydraulic conductivity occurred.

The study showed that low-pH cements can have a very moderate effect on contacting mixed-layer clays of the Friedland Ton type. This means that a concrete bulkhead wall with a thickness of several decimetres will not have any significant short- or long-term impact on a contacting backfill of dense FIM clay at near-room temperature.

7.3.8   *Performance of strongly compressed smectite-rich clay*

Background

Several attempts have been made to produce dense smectite-rich buffer clay and borehole plugs by pouring or blowing pellets of MX-80 type pellets. The density is low, however, and attempts have therefore been made to prepare very strongly compressed dense pellets to reach a sufficiently high density after hydration and homogenization. The performance of clay formed by such strongly compressed pellets has been investigated to determine if it has the same properties as clay prepared using ordinary MX-80 grains. The study also involved exposure of the pellet mass to hot vapour, simulating the actual conditions for the buffer clay. The clay powder used for pelletization had been dried to a water content of about 2%, and the pressure used in the preparation of the pellets, which were smaller than 3 mm was about 200 MPa. Their water content was 1.9–2.3% after storage under normal room conditions. This was in contrast to the range of 7–10% for ordinary MX-80 clay, and shows that the strongly compressed pellets have a very low hydration potential.

Exposure of smectite clay to hot water vapour has been found to reduce the expandability of smectite clay – as reported by Couture for his experiments with water vapour at 150–250°C [27]. He concluded that the loss of expandability is related to the influence of pressurized water vapour but gave no explanation of the mechanisms involved. We will examine a later study of strongly compacted MX-80 pellets including an experiment with hot steam.

Properties of clay prepared by compacting very strongly compressed smectite pellets

The very high density of strongly compressed clay pellets makes them useful in preparing dense backfill by blowing them into the space to be filled. However, it is necessary to ascertain that the pellets will expand to obtain a homogeneous clay with swelling pressure and conductivity similar to or approaching that of the ordinary MX-80 clay after water saturation and cooling. The following is a discussion of the study designed to determine the behaviour of the pellets [28].

Equipment and material

Oedometer cells were equipped with heater cables (Figure 7.25) to heat the samples to about 65°C before and during saturation with distilled water. After complete saturation the densities of two samples were 1660 kg/m^3 and 1930 kg/m^3, respectively. The oedometer filters were connected to burettes for continuous measurement of the amount of absorbed water.

Swelling pressure and hydraulic conductivity

The evolution of the swelling pressure is illustrated by Figure 7.26. It shows the obvious difference in swelling pressure of a clay prepared from ordinary MX-70 grains and very strongly compressed MX-80 pellets.

The hydraulic conductivity of the finally matured clay prepared using strongly compressed MX-80 pellets was two orders of magnitude higher

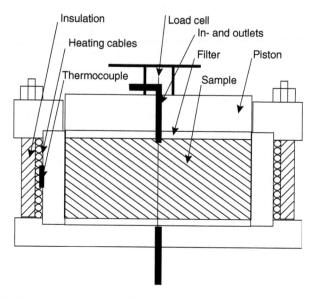

Figure 7.25 Schematic picture of cell with pellets confined during water saturation and heating. Recording of swelling pressure and hydraulic conductivity made during and after saturation.

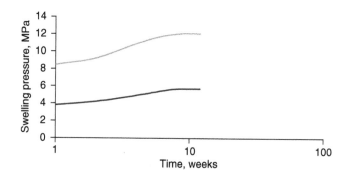

Figure 7.26 Evolution of swelling pressure of strongly compressed MX-80 pellets at room temperature. Upper curve: density at water saturation = 1980 kg/m³. Lower = 1895 kg/m³. The swelling pressure of ordinary MX-80 is about 3.5 and 1.5 MPa for these densities.

than that of ordinary MX-80 clay in the first few months after the start of oedometer testing. It then dropped to nearly but not the same value as for the ordinary MX-80 clay, indicating a significantly delayed homogenization process. This slow and apparently incomplete maturation is demonstrated in Figure 7.26.

Microstructural investigation

TEM was used to examine the microstructure of water-saturated clay – to find an explanation for the anomalous behaviour of clays prepared with the strongly compressed pellets. Small specimens were embedded in acrylate after stepwise exchange of the porewater by alcohol, using the technique described in Chapter 3. Ultrathin sections were microtomed for TEM micrograph after polymerization of the impregnated samples. Micrographs of the clay prepared with strongly compressed pellets revealed that many aggregates of stacks of lamellae had not expanded uniformly as illustrated in Figure 7.27. Strong binding resulting from interparticle action and cementation amongst and between many of the stacks of lamellae contributed to resistance against swelling.

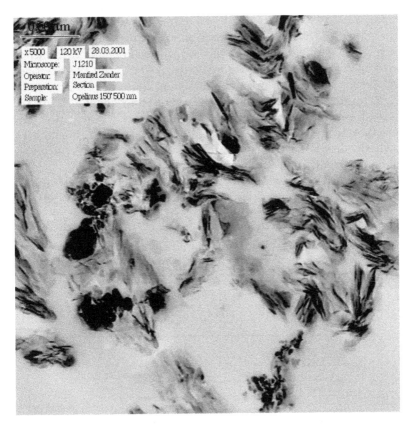

Figure 7.27 TEM micrograph of clay prepared by strongly compressed MX-80 pellets; density at saturation = 1750 kg/m³. The stacks of lamellae form dense, largely spherical aggregates and not the wavy network of stacks in ordinary MX-80 clay. Edge length of the photo is 10 μm (Photo J. Kasbohm).

*The effect of hot water vapour on clay prepared from
strongly compressed MX-80 clay pellets*

Heating during the wetting process, similar to conditions that prevail in a HLW repository, can generate thermal stresses in the pellets that would contribute to the disintegration of the pellets. This possibility, which has been investigated by Pusch, Bluemling Johnson [28], used autoclave treatment for 30 days at 110–150°C of pellets compacted to a dry bulk density of 1350 kg/m³. The pellets had an initial water content of 2.5%. Distilled water was contained in cups placed on the free upper surface of the samples in the hydrothermal cells. This amount was theoretically sufficient to allow for complete interlamellar hydration.

The difference in expandability caused by hot vapour treatment is illustrated in Figure 7.28. This figure shows the expansion and gel formation observed in laboratory experiments. A significant difference in dispersion and gel formation capacity of ordinary MX-80 powder and strongly compressed MX-80 pellets was observed. It is believed that this is the result of the combined effect of strong cohesion of clay particles brought together by the very high pressures at the time of pelletization, and by cementation caused by vapour attack. Vapour treatment at 110°C and 125°C resulted in development of swelling pressures similar to those of unheated ordinary MX-80 powder, but significantly lower, that is, only about 25%, of that of unheated strongly compressed pellets. As expected, thermally induced stresses disintegrated the clay pellets. However, vapour treatment at 150°C produced a drop in swelling pressure that can be ascribed to cementation at cooling by precipitation of silica released by the hot vapour.

Electron microscopy was used for examination of the microstructure and for element analysis, to determine whether cementing precipitates had

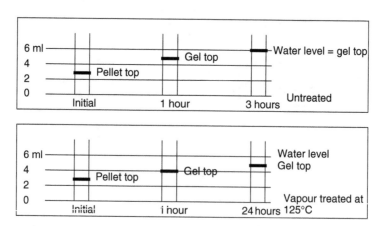

Figure 7.28 Schematic picture of the difference in dispersion and gel formation of strongly compressed MX-80 pellets without and with vapour treatment [24].

formed in the vapour tests. It was concluded that there were two major differences between vapour-treated and untreated clays prepared from strongly compressed pellets: (1) The first-mentioned had denser aggregates and larger voids, indicating that some irreversible contraction of the stacks of montmorillonite lamellae occurred because of the vapour treatment (2) siliceous cementing compounds had precipitated, reducing the swelling potential of this clay. Such precipitation was also confirmed by applying SEM with element analyses of the contacts and surfaces of aggregates. They indicated precipitated siliceous matter that was set free from the smectite and accessory minerals by the hydrothermal treatment.

7.3.9   Note on the interaction of metal waste containers and buffer clay

Mechanical impact

The high swelling pressure on waste containers can be a significant factor in repository stability. Canisters with spent reactor fuel have an internal space in the form of channels where the fuel rods with claddings are placed, and very high pressures can deform the system and possibly create critical mechanical stresses and corrosion. The copper liner of KBS-3V canisters, for example, can be squeezed by non-uniform pressure exerted by the buffer. This raises the issue of the importance of tangential distribution of the pressure and the EDZ-affected distribution of water at the rock/buffer contact. Experience from the large-scale field experiments in underground laboratories in crystalline rock show that the distribution is uniform. A significant difference exists in deposition holes and tunnels located in rock with little water-bearing fractures in comparison to rock with such fractures.

Chemical impact

The reason for equipping the KBS-3V iron canisters with a copper liner is that it is very resistant to corrosion. Oxygen that is initially contained in the buffer is soon exhausted. The oxidation film that is created is very thin and the amount of copper ions given off to the buffer is negligible. A more important threat is sulphur-bearing minerals in the buffer and sulphate ions brought in from the groundwater. The risk associated with them is minimized by selecting buffer clay and tunnel backfills with very low sulphur content.

If water enters the interior of KBS-3V canisters, which may occur if the welding of the copper liner is not perfectly tight or if corrosion permits water penetration, a number of chemical reactions between water and iron start. The results of studies conducted by SKB in Sweden and corresponding organizations in other countries, where unlined iron or steel canisters are contemplated for use, provide us with useful information. Two effects are noteworthy: (a) ion-exchange from the initial adsorbed cation in the buffer, commonly Na, to Fe. This will be the interlamellar spacing of the clay particles and hence the microstructure and (b) cementation by precipitation of Fe oxy/hydroxides and other Fe complexes. This latter observation points out the negative consequences of using Fe-smectites as buffer clay.

An Fe-rich montmorillonite with Ca as dominant adsorbed cation has been tested on various scales under laboratory conditions using, amongst others, a test set-up ('Mock-up') simulating a KBS-3V deposition hole on half-scale with unlimited access to granitic water at the outer boundary of the confined buffer. Sampling at regular intervals was made in the 2.5 year long experiment to determine the hydraulic conductivity and swelling pressure and to perform mineralogical analyses to determine whether significant changes had occurred. The clay powder used to manufacture the buffer blocks consisted of 85% montmorillonite, 10% fine quartz and 5% graphite.

The surface of the canister was maintained at 90°C throughout the test and samples were obtained regularly at different distances from the canister. Samples obtained after about 2 years showed significant dissolution of the montmorillonite as illustrated by the X-ray diffractogram in Figure 7.29.

Figure 7.30 illustrates the mineral composition of a specimen sedimented from a suspension of the most strongly affected buffer clay: Fe-montmorillonite is the dominant mineral but mixed-layer smectite/vermiculite and some kaolinite-type species are present as well.

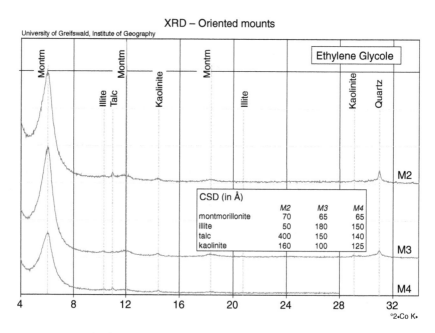

Figure 7.29 XRD plot of samples extracted at different distances from the canister in the 230 mm thick buffer after 2 years. M4 was taken from the vicinity of the canister, M2 at about 180 mm distance from it and M3 between these positions (Greifswald University, J. Kasbohm).

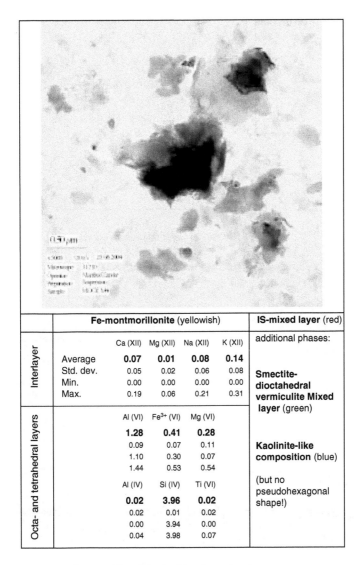

		Fe-montmorillonite (yellowish)				IS-mixed layer (red)
Interlayer		Ca (XII)	Mg (XII)	Na (XII)	K (XII)	additional phases:
	Average	**0.07**	**0.01**	**0.08**	**0.14**	
	Std. dev.	0.05	0.02	0.06	0.08	**Smectite-**
	Min.	0.00	0.00	0.00	0.00	**dioctahedral**
	Max.	0.19	0.06	0.21	0.31	**vermiculite Mixed**
Octa- and tetrahedral layers		Al (VI)	Fe³⁺ (VI)	Mg (VI)		**layer** (green)
		1.28	**0.41**	**0.28**		
		0.09	0.07	0.11		**Kaolinite-like**
		1.10	0.30	0.07		**composition** (blue)
		1.44	0.53	0.54		
		Al (IV)	Si (IV)	Ti (VI)		(but no
		0.02	**3.96**	**0.02**		pseudohexagonal
		0.02	0.01	0.02		shape!)
		0.00	3.94	0.00		
		0.04	3.98	0.07		

Figure 7.30 Mineral assembly of the buffer clay after 2 years.

The most important changes in the physical properties, as concluded from oedometer tests of samples extracted from the big Mock-up test were:

- The swelling pressure of the clay close to the canister had dropped to about 1/3 of that of untreated buffer (Figure 7.31) and to about 50% at half the distance from the canister. One reason for this was concluded to be a loss in smectite and uptake of K possibly representing a first step

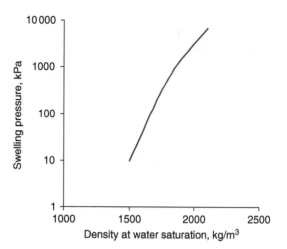

Figure 7.31 Swelling pressure of the untreated buffer clay saturated with distilled water.

Figure 7.32 Sediment volume of 11 g of moist clay (w = 29.5%) in 30 ml distilled water after 1 day. Sample M4 to the left and M1 to the right.

in illitization since Al became more abundant in the tetrahedral layers. Another reason can be precipitation of Fe-compounds formed by Fe released from the dissolved smectite.

- The mechanical strength of the buffer had increased significantly especially close to the canister, which was believed to be caused by cementation. This was supported by sedimentation tests that showed that dispersion of the most affected M4 material gave a sediment volume of about 75% of that of untreated buffer clay (Figure 7.32).

- After sectioning of the sample from the vicinity of the heater (M4), it became obvious that the clay had an aggregated nature – in contrast to the untreated material. This suggests that partial collapse of stacks of lamellae and precipitation of cementing agents had resulted as a consequence of the thermal gradient.
- The hydraulic conductivity increased by at least an order of magnitude. This was believed to be caused by microstructural rearrangement due to cation exchange from Ca to Fe.

A general conclusion is that high iron contents of buffer clays makes them less suitable as buffer material than those with a low iron content. The untreated clay used in this example had about 17% Fe^{3+}. This appears to be a proportion of Fe^{3+} that is not suitable for a buffer material in a HLW repository. However, where the temperature impact is smaller, as in borehole plugs and backfills, the material may be acceptable, provided proper experiments are conducted to ascertain the vulnerability of the iron content.

7.4   References

1. Yong, R.N., 1967. 'On the relationship between partial soil freezing and surface forces' *Physics of Snow and Ice*, H. Oura (ed.), Bunyeido Printing Co., Sapporo, Japan, 1375–1386.
2. Yong, R.N., 1965. 'Soil suction effects on partial soil freezing', *Highway Research Board*, Vol. 68: 31–42.
3. Yong, R.N., Boonsinsuk, P. and Tucker, A.E., 1984. 'A study of frost-heave mechanics of high-clay content soils'. *Journal of Energy Resources Technology*, Transactions of the American Society for Mechanical Engineering, Vol. 106: 502–508.
4. Yong, R.N., Boonsinsuk, P. and Tucker, A.E., 1986. 'Cyclic freeze–thaw influence on frost heaving pressures and thermal conductivities of high water content clays'. In *Proceedings of Fifth International Offshore Mechanics and Arctic Engineering*, Vol. 4: 277–284.
5. Anderson, D.M., Pusch, R. and Penner, E., 1978. 'Physical and thermal properties of frozen ground. C.2', in *Geotechnical Engineering for Cold Regions*, B. Andersland and D.M. Anderson (eds), MacGraw-Hill Co., New York.
6. Horseman, S.T. and Harrington, J.F., 1997. Study of Gas Generation in MX-80 Buffer Bentonite. British Geological Survey, WE197/7.
7. Pusch, R., Muurinen, A., Lehikoinen, J., Bors, J. and Eriksen, T., 1999. Microstructural and chemical parameters of bentonite as determinants of waste isolation efficiency. Final Report EUR 18950 EN, EC Contract No. F14W-CT95-0012.
8. Pusch, R., 1994. *Waste Disposal in Rock. Developments in Geotechnical Engineering*, 76. Elsevier Science Publishing Company.
9. Pusch, R., 2000. On the Risk of Liquefaction of Buffer and Backfill. SKB Technical Report TR-00-18. Swedish Nuclear Fuel and Waste Management Co. Stockholm.
10. Weaver, C.E., 1979. Geothermal Alteration of Clay Minerals and Shales: Diagenesis. Technical Report ONWI-21, ET-76-C-06-1830 Contract. Battelle–Office on Nuclear Waste Isolation.

11. Nadeau, P.H. and Bain, D.C., 1986. 'Composition of some smectites and diagenetic illitic clays and implications for their origin'. *Clays and Clay Minerals*, Vol. 34: 455–464.

12. Pusch, R. and Madsen, F.T., 1995. 'Aspects of the illitization of the Kinnekulle bentonites'. *Clays and Clay Minerals*, Vol. 43, No. 3: 133–140.

13. Pytte, A.M., 1982. The kinetics of the smectite to illite reaction in contact metamorphic shales. Thesis MA, Dartmouth College, NH.

14. Forslind, E. and Jacobsson, A., 1975. *Clay-water Systems. Water, a Comprehensive Treatise*, Vol. 5. Plenum Press, New York and London.

15. Kolarikova, I., 2004. Thermal properties of clays in relation to their use in the nuclear repository. PhD thesis Institute of Geochemistry, Mineralogy and mineral Resources, Charles University, Prague 2004.

16. Pusch, R. and Karnland, O., 1988. Hydrothermal Effects on Montmorillonite. SKB Technical Report TR 88-15. Swedish Nuclear Fuel and Waste Management AB, Stockholm.

17. Mueller-Vonmoos, M., Kahr, G., Bucher, F. and Madsen, F.T., 1990. 'Investigations of the metabentonites aimed at assessing the long-term stability of bentonites under repository conditions. Artificial Clay Barriers for High Level Radioactive Waste Repositories', *Engineering Geology*, R. Pusch (ed.) Vol. 28, Nos. 3–4: 269–280.

18. Pusch, R., 1993. Evolution of Models for Conversion of Smectite to Non-expandable Minerals. SKB Technical Report TR-93-33. Swedish Nuclear Fuel and Waste Management AB, Stockholm.

19. Grindrod, P. and Takase, H., 1994. 'Reactive chemical transport within engineered barriers', in the *Proceedings of the Fourth International Conference on Chemistry and Migration Behaviour of Actinides and Fission Products in the Geosphere*. R. Oldenburg Verlag, Muenchen.

20. Pusch, R., Takase, K. and Benbow, S., 1998 Chemical Processes Causing Cementation in Heat-affected Smectite – the Kinnekulle Bentonite. SKB Technical Report TR-98-25. Swedish Nuclear Fuel and Waste Management AB, Stockholm.

21. Petzhold, L.R., 1983. 'A description of DASSL: a differential/algebraic system solver', in *Scientific Computing*, Vol. 5: 65–68.

22. Pusch, R., 1985. Final Report of the Buffer Mass Test – Volume III: Chemical and Physical Stability of the Buffer Materials. Stripa Project Technical Report 85-14. SKB, Swedish Nuclear Fuel and Waste Management AB, Stockholm.

23. Kasbohm, J., Henning, K.-H. and Herbert, H.-J., 2000. Zu Aspekten einer Langzeitstabilität von Bentonit in hochsalinaren Lösungen. Beiträge zur Jahrestagung Zuerich 30 Aug.–1 Sept. berichte der Deutschen Ton- und Tonmineralgruppe e.V.-DTTG 2000.

24. Pusch, R. 1982. Chemical Interaction of Clay Buffer Materials and Concrete. Technical Report SFR 82-01. SKB, Swedish Nuclear Fuel and Waste Management AB, Stockholm.

25. Huertas, F., Farias, J., Griffault, L., Leguey, S., Cuevas, J., Ramirez, S., Vigil de la Villa, R., Cobena, J., Andrade, C., Alonso, M.C., Hidalgo, H., Parneix, J.C., Rassineux, F., Bouchet, A., Meunier, A., Deacarreau, S., Petit, S. and Vieillard, P. 2000. Effects of cement on clay barrier performance, ECOCLAY Project. Final Report Contract No. F14W-CT96-0032, European Commission, Brussels.

26. Pusch, R., Zwahr, H., Gerber, R. and Schomburg, J., 2003. 'Interaction of cement and smectitic clay – theory and practice'. *Applied Clay Science*, Vol. 23: 203–210.

27. Couture, R.A., 1985. 'Steam rapidly reduces the swelling capacity of bentonite'. *Nature*, Vol. 318: 50.

28. Pusch, R., Bluemling, P. and Johnson, L., 2003. 'Performance of strongly compressed MX-80 pellets under repository-like conditions'. *Applied Clay Science*, Vol. 23: 239–244.

8 Theoretical modelling of the performance of smectite clay microstructure

8.1 Introduction

In the design of engineered barrier systems of a waste landfill or a deep geological repository, modelling of the integrated performance of all the involved components is required. Currently, a number of numerical codes and analytical methods are available. We have seen in previous chapters that the evolution of the microstructure of clay seals that function as buffer systems depends on a number of factors, such as access to water and temperature. Maturation processes involving microstructural reorganization and water saturation can be described by fairly accurate conceptual and simple theoretical models. The record shows that modelling of the evolution of liners of landfill wastes and water saturation of clay seals in underground repositories can be reasonably performed for isothermal and geometrically simple 1D conditions. However, for the most critical and real practical cases of clay isolation of heat-producing high level nuclear waste (HLW), the phenomena and issues of coupled processes have yet to be fully addressed. The physical models for these coupled processes have been described in the earlier portion of this book. The record shows that existing theoretical models that include coupled thermal–hydraulic–chemical–biological (THMCB) processes have yet to meet the requirements and strict tests for verification and validation.

This chapter will focus on the clay-buffer isolation issues posed in the underground repository containment of HLW. In particular, we would be interested in how the presently available analytical-computer models (codes) fare in analysing and predicting the status of the clay buffer designed and constructed to isolate the canisters containing the HLW. The experiments and studies discussed in Chapter 7 were all designed to provide information on the properties and behaviour of the clay-buffer material under conditions evoked by the repository containment scheme. It is useful to determine how the analytical-computer codes deal with the problem at hand, and how they use the information gained from laboratory tests and studies to verify the codes. Application of the codes to predict the performance of prototype tests such as those conducted in the underground research laboratory of the Swedish Nuclear Fuel and Waste Company (SKB) at AEspoe (about 200 km south of Stockholm) would provide one with a measure of code validation. The particular prototype test which serves as a platform for the discussions detailed in this chapter originate from the EU-supported 'Prototype Repository Project'. This is a full-scale test of the KBS-3V concept of HLW isolation conducted at about a 400 m depth in crystalline rock. The same terminology used in the preceding chapters for the various clay seals will be used in the description of the repository project. These include: buffer for the dense clay surrounding the hot canisters, backfill for the less heat-affected material placed in drifts and tunnels and the less heated borehole plugs.

8.2 Purpose and objectives of models

Requirements for the kinds of analytical and computer models (ACMs) vary depending on the objectives or purpose to which the models would be applied or used. Three broad categories can be identified as follows.

Forensic – common tool used by regulatory and enforcement agencies to determine culpability or responsibility for the situation being investigated. Using site contamination by pollutants as an example – forensic models are used to determine source of pollutants and transport history. More often than not, available information concerning initial properties, boundary conditions and other details dealing with sources, loading are not known or well defined. The parameters, factors and processes needed to develop a workable forensic computational model are likely to be limited.

Performance assessment – models developed to describe system performance are most often used to assess performance of a specific system. Performance assessment models (PAMs) are also used to assist in the management of the system, and as monitoring tools to assist in aspects of regulatory and enforcement assessment. In contrast to forensic models, we note that for simple situations and systems, initial and boundary conditions for PAMs can be well identified. Additionally, for the simple systems, the processes contributing to overall system behaviour are also known. For complex systems where many interacting processes and couplings are present, and where boundary conditions are not fully understood or identified, the accuracy of analyses or results produced from application of PAMs can be suspect.

Much depends on the capability of the individual components or submodels of the PAM to reflect individual and coupled processes, and with the system model itself to accord with problem conceptualization. For those who seek accurate performance assessments of system behaviour, it is axiomatic to state that a correctly posed system problem is mandatory in the construction of the PAM. The various processes that combine to render the system operative and functional must be correctly portrayed in the model.

Prediction of system behaviour – these models are often similar to PAMs in overall organic structure. When used as a design aid, prediction models do not have the detailed operating or performance information used to construct PAMs. Prediction models developed after system operation are often direct off-shoots of PAMs, and may indeed be PAMs used for prediction purposes or offshoots thereof.

It is not the purpose of this chapter to enter into a discussion of the use of ACMs as such. The material presented in this chapter is meant to provide: (a) a brief overview of the THMCB modelling challenge, as it relates to the isolation of HLWs in repositories and (b) a summary account of some of the models developed to address the THMCB repository problem – together with some examples of their application.

8.3   Elements of the THMCB problem

The THMCB repository problem has been described in the previous chapters. We will define and examine the various elements of the problem. This can be done by grouping them into two convenient groups as: (a) *Problem initialization* and (b) *Operational*.

The basic elements of Problem initialization include the following:

- long-lived heat source
- smectite-based buffer barrier or containment system
- surrounding host rock at a temperatures much lower than the heat source
- access to water containing dissolved solutes and delivered via fractures and fissures in the host rock to the cooler extremities of the buffer
- interface conditions at the heat source end (source-buffer interface) at the rock end (buffer-rock interface) and at the top (buffer-backfill interface)
- thermodynamic potentials.

Not all the elements (processes) in the Operational group are well identified or fully understood. The ones that are known include the following:

- Heat and mass transfer
- Vapourization and condensation
- Vapour and liquid transport
- Accessibility and distribution of water intake
- Transport and transmissibility coefficients
- Swelling of buffer material on water uptake
- Swelling and compression pressures
- Solute transport
- Chemical reactions and transformations

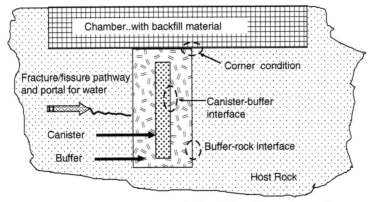

Figure 8.1 The simplified THMCB repository problem.

- Biotic reactions and transformations
- Transformed mechanical, transmission and transport coefficients
- Couplings between heat, mass, chemical and biological phenomena.

The simple schematic shown in Figure 8.1 portrays the THMCB repository problem and highlights some of the elements described in Problem initialization and Operations. The nature of the physical contact between buffer material and host rock, and also with canister and the upper backfill material is particularly important. These contacts control both the mechanical performance aspects of the buffer and the distribution of water access to the buffer material. The various processes associated with thermal, fluid, vapour, solute and gas transport, geochemical reactions, and abiotic and biotic reactions all lead to maturation of the buffer material. In the previous chapters, the results of maturation of the buffer material have been called the *evolution of the clay buffer* or *microstructural evolution*.

8.4 Basis for modelling

8.4.1 General

As we have stated in previous chapters, a basic condition for development and production of ACMs is the truthful and accurate replication or representation of the many individual and coupled processes and final system process of the target system problem. Recognizing that the factors and processes

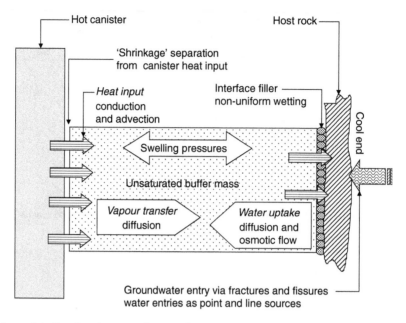

Figure 8.2 Simple schematic showing basic elements of the canister-buffer problem to be modelled.

shown in Figure 8.2 are by no means detailed or complete, it is evident that for an accurate portrayal of the many processes involved, it is necessary to obtain a good understanding of the physical, chemical and biological aspects of the problem at hand. Since the canister-isolation problem is located in a geologic environment, these various aspects translate into geophysical, geochemical and biogeochemical phenomena.

ACMs are, in essence, approximations of reality. The more complex the processes that combine to produce final system behaviour, the more one needs to be careful that all model representations of individual component behaviour that combine to form the final system model are more than simple crude approximations. The phenomenon of heat transfer or movement is probably the simplest to envisage in a simplified process-schematic form. This is shown in Figure 8.2.

Water movement in the buffer material occurs as a result of the thermal gradient generated by the two different temperatures at the opposing ends of the buffer material. Since vapour formation occurs, total transfer of water will include both liquid water and vapour. However, there are some reinforcing and counter-opposing moisture fluxes that enter into the overall water movement problem. These can be complicating factors because separation of the individual contributions from each process involved is not easily accomplished. Swelling of the constrained buffer material in the presence of water will produce resultant swelling pressures and local compression of the material. Density and moisture content changes are inevitable. The importance in obtaining accurate information on the transient state of moisture in the buffer is evident.

8.4.2 *Liquid water movement and temperature gradients*

The transient heat and mass transfer phenomena are immediately evident from the schematic diagram shown in Figure 8.2. The previous chapters and discussions have shown that the five possible reasons for water movement in the liquid phase in an unsaturated soil under a thermal gradient include:

- Surface tension gradients – Since the surface tension of water against air decreases as the temperature increases, liquid water will flow from the warm regions to the colder or less warm regions because of the influence of surface tension gradients. This has sometimes been referred to as the thermo-capillary movement, and is akin to the Gibbs Marangoni effect.
- Temperature influence on soil-water potential ψ – The soil–water potential ψ is a function of both temperature and water content, and can be expressed in a generalized way as:

$$\frac{\partial \psi}{\partial x} = \frac{\partial \psi}{\partial T}\frac{\partial T}{\partial \psi} + \frac{\partial \psi}{\partial \theta}\frac{\partial \theta}{\psi}$$

where x, T and θ denote spatial coordinate, temperature and volumetric water content respectively. For the same volumetric water content, the soil–water potential increases (becomes more negative) as the temperature increases.
- Flow of water from the less warm regions to the warmer regions because of differences in the respective heat contents between the liquid water layers sorbed onto the solid particle surfaces and the bulk liquid in the pore spaces. In a sense, this is sometimes interpreted as thermo-osmotic movement.
- Water movement generated by the changes in the random kinetic energy of the hydrogen bonds.
- Ludwig–Soret effect. This cross effect between temperature and concentration results in the phenomenon of thermal diffusion, as for example in the diffusion of solutes in the pore water from the warm regions to the less warm regions.

8.4.3 *Vapour movement*

Whilst diffusion of vapour in the buffer mass is generally viewed as a Fickian process, the work reported by Smith [1] on the thermal transfer of water in unsaturated soils has argued that this is not strictly true. To a very large extent, this is because of the rapid condensation of vapour at one end and the evaporation of water at the other end. Reports by various soil scientists on work performed as early as the 1950s, cf. [2, 3, 4] confirmed: (a) that circulation of liquid and vapour occurs in closed systems of porous materials and (b) that vapour movement is from four to five times greater than that expected by diffusion according to the Fickian model.

8.4.4   Swelling and compression

In a constrained environment, swelling of a buffer mass consisting of swelling soil materials upon water entry and uptake will produce compressive forces throughout the wetted portion. These will be transmitted to the unwetted portion of buffer mass at the wetting front. Progressive compression of the unwetted portion of the buffer mass will result. The simplified conceptual experiment shown in Figure 8.3 illustrates the phenomenon.

The horizontal column shown at the top of the figure portrays the initial condition ($t = 0$) where macroscopically, the specimen is of uniform density, porosity and water content. The uniform light shading shown in the specimen is meant to convey the macroscopic uniform conditions. From a microstructure perspective, however, the specimen is not uniform in any of its properties. The samples on the right (bottom half of picture) are ideal specimens with no microstructure, and hence exhibit total uniformity in density and porosity. The specimens on the left are controlled by the

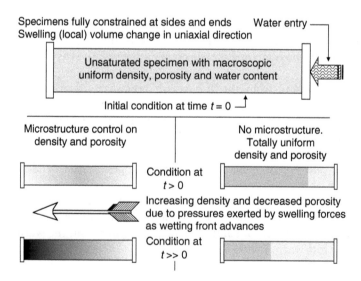

Figure 8.3 Illustration of a laboratory experiment demonstrating the effect of swelling of buffer material in a constrained sample. Note the initial macroscopic uniform condition prior to water entry into the sample. Increased darkness in shading illustrates increasing density of sample.

presence of microstructures, and hence do not exhibit local uniformity in density or porosity. As water enters into the specimen, that is, as water uptake occurs, the wet front progresses from right to left and buffer soil swelling occurs. This is the condition at some time $t > 0$. Since the specimen is fully constrained, the swelling pressures developed in the wetted portion would react against the constraints. The unwetted portion of the specimen at the wetting front forms one-end constraint.

For the specimens with no microstructure, uniform compression of the unwetted portion is obtained as a result of swelling of the wetted portion. The results are shown on the bottom right with uniform shadings of light and dark gray. For the samples with microstructure control on macrostructure and properties, non-uniform compression of the unwetted portion occurs because of: (a) the microstructure of the buffer material (b) the diffusive movement of water ahead of the wetting front and (c) variation of water contents ahead and behind the wetting front.

The compressive forces resulting from the constrained swelling performance of the buffer material will be directly related to at least two factors: (a) the amount of actual volume change occurring in the wetted portion and (b) the rigidity and degree of uniformity of the unwetted portion. The two factors are in actuality one single complex phenomenon of continuous adjustment of compressive and expansive performances. If one assumes complete rigidity of the unwetted portion, there will be no expansion or volume change in the wetted portion. The consequence of this assumption are: (i) total swelling pressures, that is, compressive forces will be developed in accord with the initial particle spacing of the material in place since the underlying tenet is that there is no microstructural control, and that all particles act individually and are dispersed uniformly (ii) the total swelling pressures in the wetted portion are uniformly distributed, and these are also the same pressures transmitted via the rigid unwetted portion to the other end of the column specimen shown in Figure 8.4. In short, the result of this assumption is that the compressive forces resulting from the constrained swelling will be constant and essentially not dependent on the amount of water uptake.

One can argue that a quick examination of the schematic shown in Figure 8.4 will show that the magnitude of the double-ended arrows shown are all equal and constant so long as saturation of the wetted portion is obtained. Furthermore, if this saturation is attained, the amount of advance of the wetting front does not impact on the pressures and forces generated as these are constant. This is a consequence of the no-volume change condition. It is evident that the rigid block assumption is perhaps too extreme an assumption.

The case for assuming some compressibility of the unwetted portion of the buffer mass automatically or implicitly recognizes that: (a) volume change will occur in the wetted portion upon water uptake, (b) microscopic non-uniformity exists in the buffer mass. This is the control exercised by the

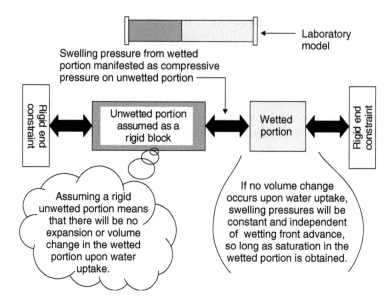

Figure 8.4 Assumption of properties and behaviour of unwetted portion as a rigid block in the face of wetting of the buffer mass.

presence of microstructures and their distribution and/or arrangement in the buffer mass and (c) compression is not uniform in the unwetted portion because of non-uniformity in porosity and density. Figure 8.5 shows a schematic of the laboratory experiment previously discussed. The unwetted portion is fragmented in *n* fragments to show differences in density and porosity in the unwetted portion.

Unequal longitudinal compression occurs in the unwetted portion of the buffer mass because the compressible nature of the material is controlled by the microstructural non-uniformity of the specimen. This will vary as the wetting front advances, thus requiring one to develop relationships for constitutive or rheological performance that recognize the wetted portion performance characteristics. In the transient state, since complete wetting of the buffer mass in real life can take hundreds and thousands of years, stress equilibrium and density equilibrium will not be attained. When full equilibrium is attained, one would expect macroscopic uniform densities and water contents.

8.4.5   *Chemical and biological issues*

There are a number of chemical phenomena that will impact directly on the performance of the buffer mass and hence on the analytical-computer modelling problem itself. Two particular phenomena are worthy of mention at

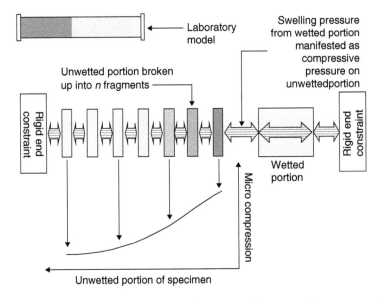

Figure 8.5 Schematic showing compression of unwetted portion of specimen resulting from compression forces developed from swelling of the wetted portion. The difference in shading of the *n* fragments of the specimen is meant to portray differences in density and porosity.

this stage. The two broad categories include: (a) transport of dissolved solutes due to the presence of thermodynamic gradients provoked by T and H, and their effects on the THM performance characteristics and (b) transformation of the clay materials in the buffer mass due to T, H, C and B and the resultant effects on the performance characteristics of the buffer mass.

At the present time, most models are concerned with the prediction of transport and fate of contaminants, as for example the very popular dispersion–diffusion relationship given in simple form as:

$$R\frac{\partial c}{\partial t} = D_L\frac{\partial^2 c}{\partial z^2} - v\frac{\partial c}{\partial z} \tag{8.1}$$

where $R = [1 + (\rho/n\rho_l)k_d]$ is the retardation coefficient or factor, c = concentration of target solute, D_L is the longitudinal dispersion–diffusion coefficient, z is the spatial coordinate, v is the advective velocity, ρ is the bulk density of soil media, ρ_l is the density of pore liquid, k_d is the partition coefficient, and n is the porosity of soil media.

The relationship given as Eq. (8.1) presumes that one can obtain a representative partition coefficient k_d for the particular target solute and for the soil media under investigation. Note that if k_d is assumed to be a

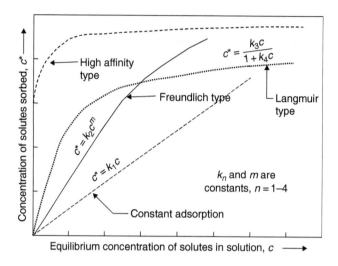

Figure. 8.6 Different types of adsorption isotherms obtained from batch equilibrium tests. Note c = concentration of target solutes and c^* = concentration of target solutes sorbed by soil solids.

constant, this means that the partitioning of solutes onto the soil solids will be a linear function, that is, that we will have infinite sorption of solutes by the soil solids as seen in Figure 8.6 where the classical adsorption isotherms are portrayed. The slope of the constant adsorption line is given as k_d. Evidence from countless batch equilibrium adsorption isotherm tests has shown adsorption phenomena approaching the Freundlich-type or Langmuir-type of sorption behaviour. The reader is advised to consult the textbooks specializing in the transport and fate of pollutants for more detailed information [4, 5].

The widely used standard dispersion–diffusion equation given as Eq. (8.1) is not a reactive transport relationship, that is, it pays no attention to resultant reactions, complexations and speciations that occur when the solutes interact with other solutes and ligands in the pore fluid and the reactive surfaces, and especially with the ions in the interlayers of the soil solids. It is important to take note that in respect to the transport of dissolved solutes in the saturated and unsaturated buffer material, it is not the transport phenomenon by itself that merits total attention. It is the effect and influence of the increased or decreased concentrations and species of solutes on the transmissivity and mechanical properties and performance of the buffer material that are of paramount importance. Section 8.4.6 discusses some of the more pertinent and important aspects in relation to the microstructure of the buffer material.

Determination of the distribution of multi-species solutes in the buffer in relation to time requires one to systematically solve individual and multi-species transport relationships in a chemically reactive mode. The simplest approach is to consider the chemical reactions between solutes and ligands in terms of the resultant effect on the chemical potential ψ_c. Accordingly, we can express the effects of temperature and moisture content on the transport of dissolved solutes in a partly saturated buffer mass, as a first approximation, in the form of a simple extended relationship as follows:

$$\frac{\partial \theta c}{\partial t} = \frac{\partial}{\partial x}\left[k(\theta, c)\frac{\partial \psi_c}{\partial x}\right] - k(\theta, c)\left[\frac{\partial \psi(\theta, T)}{\partial x} \frac{\partial c}{\partial x}\right] - \frac{\partial (\rho^* c^*)}{\partial t} \tag{8.2}$$

where c is the concentration of target solute, ψ_c is the chemical potential is $\psi(\theta, T)$, ρ^* is the bulk density of buffer material divided by the bulk density of water and c^* is the concentration of target solute sorbed onto the solid buffer materials.

Since both k and ψ are dependent on θ and c or T, a solution to Eq. (8.2) will not be easily obtained without recourse to other governing relationships. Considering T as the absolute temperature, P and m as the pressure and mass of the buffer material and considering these as independently chosen variables, the thermodynamic potential (i.e. the chemical potential ψ_c in this case) can be obtained in terms of partial molar free energies. With the condition that at equilibrium, G^* the partial molar free energy is the same everywhere. Differentiation of the Gibbs energy relationship in terms of the partial molar free energy and in the absence of significant gravitational and external forces, will yield:

$$dG_\omega^* = V_\omega^* dP - S_\omega^* dT + \frac{\partial G_\omega^*}{\partial \theta} d\theta \tag{8.3}$$

where G_ω^* is the partial molar free energy of the soil-water and is considered also to be equal to the chemical potential ψ_c, and V_ω^* and S_ω^* represent the partial specific volume and partial specific entropy of the soil-water respectively. By definition, $V_\omega^* = 0$ for a soil that does not shrink or swell upon water content changes and $V_\omega^* = 1/\rho_w$ when the soil is fully saturated (ρ_w = mass density of water). Note that the soil-water potential $\psi = \psi_c - \psi_c^o$, and that ψ_c^o = standard state for ψ_c.

Transformations in the buffer material due to temperature (T), presence of water (H) and biological activity (B) occur over long time periods. Transformation of clay minerals is sensitive to the influence of temperature and water availability. Section 7.3 has described some of the transformation mechanisms and results – that is, illitization and precipitation of silica. Previous studies have focussed on transformations of minerals due to weathering forces not unlike the situation at hand, that is, temperature and presence of water. Figure 8.7 is a pictorial representation of a proposed weathering scheme.

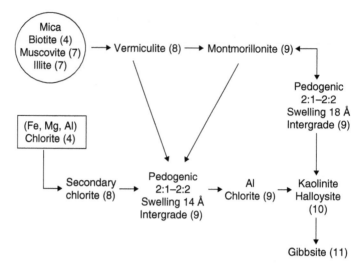

Figure 8.7 Transformation of various clay minerals. Numbers in parentheses are weathering indices with more stable minerals having higher numbers.

The chemical reactions that promote dissolution and precipitation, depending on the type of clay mineral, are greatest at higher temperatures and higher moisture contents. Water is both a protophilic and a protogenic solvent. Since it can act either as an acid or as a base, it is amphiprotic. It can undergo self-ionization (autoprotolysis), resulting thereby in the production of a conjugate base OH^- and a conjugate acid H_3O^+.

Microbial species detected in the groundwater in deep vault repositories include aerobic and microaerophilic heterotrophic microorganisms, and anaerobic iron-reducing and sulphate-reducing bacteria. The importance of recognition of the capability of microorganisms such as facultative anaerobic bacteria, fungi and even anaerobes in reducing iron cannot be overstated. The presence of water in the buffer material allows both acid-base reactions and also oxidation–reduction reactions. The latter can be abiotic and/or biotic. In respect to biological issues, microorganisms are significant participants in catalysing redox reactions. The activity of the electron e^- is a significant factor in oxidation–reduction reactions since these involve the transfer of electrons between the reactants. In the schematic diagram shown in Figure 8.8, the reduction of the structural Fe^{3+} in the octahedral and tetrahedral sheets to Fe^{2+} will significantly alter the short-range forces between the lamellae and could result in: (a) a lower specific surface area, (b) a high degree of layer collapse, (c) decreased water-holding capacity, (d) reduced swelling capability and (e) stacking of the layer sheets (cf. Section 7.3.9).

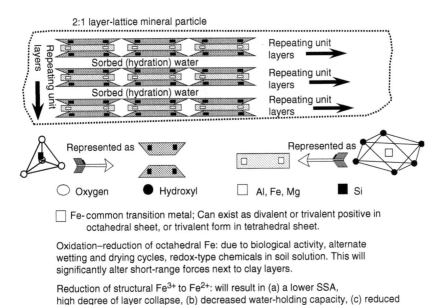

Figure 8.8 of the repeating layers of a 2:1 layer-lattice mineral particle:

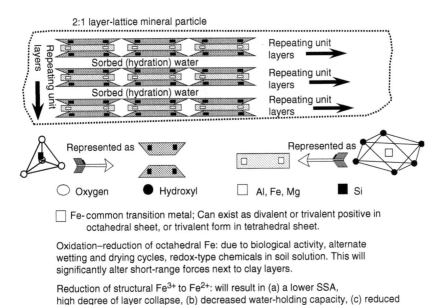

2:1 layer-lattice mineral particle

Sorbed (hydration) water

Repeating unit layers

Represented as

Oxygen   Hydroxyl   Al, Fe, Mg   Si

Fe- common transition metal; Can exist as divalent or trivalent positive in octahedral sheet, or trivalent form in tetrahedral sheet.

Oxidation–reduction of octahedral Fe: due to biological activity, alternate wetting and drying cycles, redox-type chemicals in soil solution. This will significantly alter short-range forces next to clay layers.

Reduction of structural Fe^{3+} to Fe^{2+}: will result in (a) a lower SSA, high degree of layer collapse, (b) decreased water-holding capacity, (c) reduced swelling and (d) better stacking of particles.

Figure 8.8 Schematic showing the repeating layers of a 2:1 layer-lattice mineral particle (montmorillonite) and what might happen in a reducing environment. Reduction of the structural iron (Fe) in the mineral comprising the buffer material is most likely.

8.4.6 *Microstructure controls on transmissivity and swelling properties*

To fully appreciate the influence or role of microstructural units (MUs) on the macroscopic or *bulk* transmissivity and mechanical (swelling) properties of the buffer material, we need to go back to basic considerations of inter-layer phenomena and water uptake. Figure 8.8 schematic shows the repeating layers of a 2:1 layer-lattice smectite mineral with interlayer spaces occupied by exchangeable ions. The basal spacing $d(001)$ of 0.95–1.0 nm for the anhydrous smectite (i.e. totally dry state) will expand from 1.25 nm to 1.92 nm depending on the amount of water intake (hydration). For most exchangeable cations in the interlayer space, except for Li and Na, the basal spacing expansion for montmorillonites appear to reach a maximum value of about 1.92 nm. This is about the thickness of four layers of water (i.e. four hydrate layers). The size of a unit particle of montmorillonite varies according to the exchangeable cation in the interlayer. Figure 8.9 gives a schematic depiction of MUs in a typical (idealized) unit elementary volume.

The relationship between the activities and energies in the micropores and the macropores is a very interesting one. Water uptake into an anhydrous

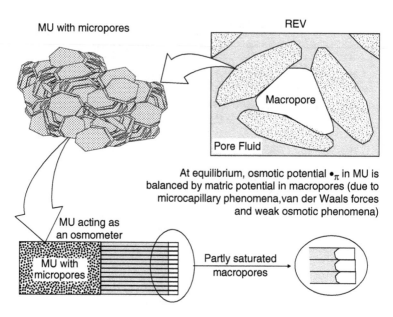

MU with micropores

REV

Macropore

Pore Fluid

At equilibrium, osmotic potential $\bullet_\pi$ in MU is
balanced by matric potential in macropores (due to
microcapillary phenomena, van der Waals forces
and weak osmotic phenomena)

MU acting as
an osmometer

MU with
micropores

Partly saturated
macropores

Figure 8.9 Schematic diagram showing the action of MUs in a representative unit
volume. Uptake of water by a typical MU is initially via hydration
forces, and later on by DDL forces.

smectite particle is via hydration forces initially and subsequently via
diffuse double-layer (DDL) forces. The energy characteristics defined by the
soil-water potentials provides an insight into how strongly water is held to
the soil solids, that is, how strongly water is held in the buffer material (in
the THM problem). Determinations in respect to the osmotic potential
and the matric potential have been found to be useful in analyses regarding
swelling pressures and volume change. Figure 8.9 shows that at final equilibrium, the osmotic or solute potential ψ_π of the MU should be equal to the
matric potential ψ_m measured as a macro property. This matric potential is
a characteristic of the macropores and is developed by the microcapillaries,
van der Waals forces, weak osmotic phenomena and other surface active
forces existent in the macropores.

The sequence of water uptake in a swelling soil depends on the nature of
the exchangeable cations in the interlayer spaces and the initial water content of the partly saturated soil. Upon first exposure to water (or water
vapour), hydration processes dominate and water sorption is an interlayer
phenomenon. For swelling soils containing Li or Na as the exchangeable
cation, continued uptake beyond hydration water status occurs due to
double-layer forces [5]. The solvation shell surrounding small monovalent
cations consists of about 6 water molecules if the solution is dilute [6]
reducing to about 3 for concentrated solutions. No secondary solvation

shells are associated with added water intake. Water uptake beyond inter-layer separation distances of about 1.2 nm occurs because of double-layer swelling forces, resulting in the formation of a solution containing dispersed single unit layers. The osmotic potential ψ_π is often used to determine the energy status and the potential for water uptake. In the case of divalent cations, both primary and secondary solvation shells move together as a solvation complex. The primary shell is seen to be composed of about 6 to 8 molecules and the secondary shell contains about 15 water molecules [6].

Initial water uptake by nearly anhydrous clays is strongly exothermic, with the water being firmly held in the coordination sphere of the cation and in contact with the surface oxygens [5]. Interlayer or interlamellar expansion due to sorption of water (hydration) is determined by the layer charge, inter-layer cations, properties of adsorbed liquid and particle size. Volume changes as high as 100% of the original volume of the dry clay can be obtained when four monolayers of water enter between the layers of a montmorillonite clay. Expansion of interlayer distance beyond hydration expansion which is ascribed to DDL mechanisms will increase volume far beyond the 100% obtained from the four monolayer hydration. This will cause the morphology of the microstructures to change. Figure 8.10 shows the swelling of a natural low-swelling clay (montmorillonite content of about 6%) with initial natural density of 1960 kg/m^3 in a wetting column experiment. The numbers next to the water content points represent the bulk density of the soil at that position. The results indicate local expansion (volume change) of the soil at the front

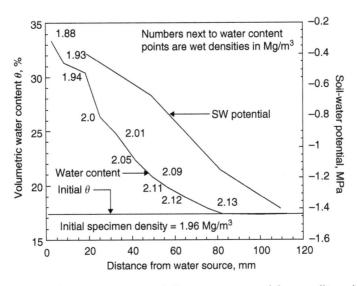

Figure 8.10 Results from unsaturated flow into a natural low-swelling clay soil column, showing volumetric water content development along the soil column and corresponding soil-water potential ψ. Water uptake is from the left.

end of the column. Theoretical calculations show that at full saturation, with a no-volume change condition, the saturated bulk density of the soil would be about 2020 kg/m^3. The corresponding soil-water potentials developed in the soil column are shown on the right-hand ordinate.

It is not unreasonable to expect significant changes in volume and shape of the microstructures as they uptake water via DDL mechanisms. The combination of increased volumes of the MUs and the interactions of the MUs during water uptake is seen as the swelling of the buffer material. Constraints to swelling will be demonstrated as resistance to swelling and is commonly described as *swelling pressure*. In THM modelling scenarios, these swelling pressures developed upon water uptake are considered as *compressive forces or pressures*. When such is done, it is necessary to account for the development of these *compressive pressures* in terms of water-uptake performance relationships.

8.4.7   Couplings, dependencies and interdependencies

Given a heat source and water availability, tracking the movement of heat and water in a smectite buffer mass requires one to be fully aware not only of the various coupled relationships between heat and fluid fluxes, but also of the dependent and interdependent relationships between these and the various internal sets of chemical and mechanical forces and events generated. Water movement in response to a thermal gradient is well appreciated, and a simple Fickian-type relationship can be written to express this phenomenon – in the absence of all other factors, forces, processes and events. However, as noted in the previous chapters, changes in water content θ in a smectite clay buffer in the repository system are due not only to thermal gradients, but also to the various internal gradients. The effects and outcome of the actions generated for many of these are inter-related and interdependent. A brief discussion of some of the more prominent factors or events that deserve attention is as follows:

- water content changes $\Delta\theta$ at any one point i in the clay-buffer system arise from the thermal gradient, evaporation and condensation, and internal energy gradients determined in terms of $\Delta\psi$;
- swelling of the smectite clay at any one point i as a function of the ambient T, θ, concentration of solutes c, compressibility of the unwetted portion;
- changes in internal energy characterized by the total potential ψ, at any one point i in the smectite clay barrier as a function of the ambient θ, T and c.

A good example of the interdependencies can be gained by looking at the simple problem of determination of the concentration of solutes in the pore-water. The concentration of solutes at any one point is considered important since this: (a) has an effect on the osmotic (swelling) pressure, (b) affects the vapour pressure of the porewater and (c) participates strongly in reactions with other solutes and reactive surfaces. Experiments on the contributions from vapour and thermo-capillary transfer of liquid water under a temperature gradient [2] for a closed soil-column system showed that whereas total (plus–minus) transfer of liquid water was towards the cold end, movement of dissolved solutes was in the opposite direction. Vapour movement was found to be four to five times greater than that predicted by Fickian diffusion.

A useful technique to account for coupled effects is to use the principles of irreversible thermodynamics to cast the governing relationships. This allows one to recognize that the situation under consideration is in a non-equilibrium state, and thermodynamic forces can be prescribed to represent the constituent components as they move towards the establishment of equilibrium – through the balance of forces. Taking the simple case of heat and mass transfer in the unsaturated buffer mass, the second postulate of irreversible thermodynamics allows one to write the governing relations as follows:

$$
\left\{ \begin{matrix} Q_w \\ Q_T \end{matrix} \right\} = \begin{bmatrix} L_{ww} & L_{wT} \\ L_{Tw} & L_{TT} \end{bmatrix} \left\{ \begin{matrix} \dfrac{\partial \theta}{\partial x} \\ \dfrac{\partial T}{\partial x} \end{matrix} \right\} \tag{8.4}
$$

where $Q_w = \partial \theta / \partial t$ = fluid flux, $Q_T = \partial T / \partial t$ = heat flux. $\partial \theta / \partial x$ and $\partial T / \partial x$ are the thermodynamic forces due to water content gradient and temperature gradient, respectively. The phenomenological coefficients are described as follows: L_{ww} is the diffusion coefficient for fluid flow due to gradient of θ, L_{TT} is the thermal conductivity coefficient for heat transfer due to gradient of T, L_{wT} and L_{Tw} are the coupling coefficients. Mass and energy conservation relationships can be specified as follows:

$$
\frac{\partial \theta}{\partial t} = -\frac{\partial Q_w}{\partial x} \quad \text{and} \quad C_v \frac{\partial T}{\partial t} = -\frac{\partial Q_T}{\partial x} \tag{8.5}
$$

where C_v = specific heat of the buffer material. Accordingly, the governing relationships will be obtained as follows:

$$
\left\{ \begin{matrix} \dfrac{\partial \theta}{\partial t} \\ \dfrac{\partial T}{\partial t} \end{matrix} \right\} = \begin{bmatrix} 1 & 0 \\ 0 & \dfrac{1}{C_v} \end{bmatrix} \frac{\partial}{\partial x} \left\{ \begin{bmatrix} L_{ww} & L_{wT} \\ L_{Tw} & L_{TT} \end{bmatrix} \left\{ \begin{matrix} \dfrac{\partial \theta}{\partial t} \\ \dfrac{\partial T}{\partial t} \end{matrix} \right\} \right\} \tag{8.6}
$$

These governing relationships have been used previously [7] to determine the phenomenological coefficients with the aid of an identification technique. The procedure requires one to assume that the phenomenological coefficients are certain types of functions of θ and T. This obviously requires an accumulated knowledge of transport of fluids in unsaturated materials. The constant parameters associated with the functional forms can be evaluated using the identification technique, and the phenomenological coefficients determined as certain functions of θ and T for the soil material.

8.5   Modelling of microstructural evolution

8.5.1   *General*

It is obvious that a number of coupled physical and chemical processes combine to give the clay a state that is transient and that it will never reach a perfectly stable condition. From the point of safety analysis, this means that the successively altered clay must fulfil certain requirements with respect to the isolation capacity, which, for HLW, is expressed in terms of allowed concentrations of released radionuclides in the groundwater and maximum allowed dose rates to which individuals or groups of people on the ground surface are exposed. The fact that the microstructure is a determinant of most of the important physical properties has led to attempts to model the microstructural evolution.

Conceptual and simplified theoretical models for the maturation of the clay microstructure have been described in Chapter 4, starting from the preparation of the various materials by compaction to the temporary state of complete fluid saturation and redistribution of clay particles. They can be further developed to include the impact of consolidation or expansion and also the effect of cation exchange and temperature. The stochastic nature of the system of particles and porewater however could present some difficulty in structuring the governing relationships. In recognition of this, an approach used by many modellers is to consider the system to be a homogeneous, isotropic porous medium with bulk properties that can be determined using traditional laboratory techniques, and to interpret and explain the results with respect to known or assumed processes on the microstructural scale. This has led to shortcomings and in some cases completely misleading results, forcing the modellers to introduce repair coefficients and to derive parameters backward from true experiments.

Theoretical models that can successfully predict performance of clay seals over thousands to hundreds of thousands of years are needed if safe management and disposal of HLW is to be achieved. Some of the models that are being used in the current work for developing safe techniques for disposal of HLW will be examined. The specific interest in these models is with respect to how well they can predict the status of the clay buffer in situations determined in the prototype repository project. Whereas the agreement or lack of agreement between predicted and experimental values can serve as a means to determine the viability of a model, the mechanistic approach taken makes it all the more important to determine why a particular succeeded or failed.

8.6 Integrated THMC modelling

8.6.1 *Introduction*

The general nature of the processes involved in the evolution of the buffer and backfill in a KBS-3V repository has been described in Section 6.3 and a conceptual model for it has been explained in detail in the first part of this chapter. The corresponding theoretical models, termed codes because of their use in numerical calculations, have been developed by international research groups working for SKB and corresponding organizations in several EU countries. These models and codes are presently being used in the international Prototype Repository Project at AEspoe to predict and evaluate major physical and chemical processes. Although the intent of the exercise is to model the behaviour of the buffer and backfill in relation to temperature (T), water migration (H), stress and strain (M), as well as to chemistry (C) and biology (B) of the system, the models developed, and used in the exercise, have not reached the level of sophistication that incorporates C and B. For the basic model that has made the attempt to include C, the non-chemically reactive transport analysis with capability for coupling with a geochemical speciation model has yet to reach the stage where these can be directly superposed onto the existing THM platform.

By and large, modelling of system performance based on a mechanistic approach requires one to accurately portray the various processes that contribute to the performance of the system. In addition, proper representations of the initial properties of the system material and boundary conditions are required if predictions of system performance are to be made. For the THMC models developed to predict performance of the engineered buffer system (EBS) in the Repository, it is essential that a means be found to provide the basis for model validation. The EU-supported 'Prototype Repository Project' mentioned in Section 8.1, that is, a full-scale test of the KBS-3V concept at about 400 m depth in crystalline rock, provides such an opportunity. All the pertinent parameters and data relating to EBS performance have been made available for THMC validation. It is recognized that for calibration purposes, a few sets of information (data) were used as control data. For modelling purposes, it is understood that the data are specific to the type of buffer material used, the specific initial and boundary conditions, and the specific test conditions imposed. Critical information regarding chemical speciation, evolution of microstructure, material transformations, and microbial flora and population and other microorganisms will be needed for further development and validation work to test the capability of the models when they address the complete package of THMCB activity. Some of the more popular codes presently in use are discussed in the following sections. These have been used by the developers or promoters of these codes in predicting the performance of the buffer material in the prototype repository experiment. The details concerning the structure and computational techniques used, together with development of the code itself are on record in the various reports written for the prototype repository project. Some of these are summarized in the following section. Prominent amongst these codes are the COMPASS, ABAQUS, BRIGHT and THAMES.

8.6.2   Code 'COMPASS'

COMPASS is based on a mechanistic theoretical formulation where the various aspects of soil behaviour under consideration are included in an additive manner. In this way the approach adopted describes heat transfer, moisture migration, solute transport and air transfer in the material, coupled with stress/strain behaviour [8].

General description of formulation employed

Partly saturated soil is considered as a three-phase porous medium consisting of solid, liquid and gas. The liquid phase is considered to be porewater containing multiple chemical solutes and the gas phase as pore air. A set of coupled governing differential equations can be developed to describe the flow and deformation behaviour of the soil.

The main features of the formulation are described below:

- Moisture flow includes both liquid and vapour flow. Liquid flow is assumed to be described by a generalized Darcy's Law, whereas vapour transfer is represented by a modified Philip-de Vries approach.
- Heat transfer includes conduction, convection and latent heat of vaporization transfer in the vapour phase.
- Flow of dry air due to the bulk flow of air arising from an air pressure gradient and dissolved air in the liquid phase is considered. The bulk flow of air is again represented by the use of a generalized Darcy's Law. Henry's Law is employed to calculate the quantity of dissolved air and its flow is coupled to the flow of pore liquid.
- Deformation effects are included via either a non-linear elastic, state surface approach or an elasto-plastic formulation. In both cases deformation is taken to be dependent on suction, stress and temperature changes.
- Chemical solute transport for multi-chemical species includes diffusion, dispersion and accumulation from reactions due to the sorption process.

Heat transfer

Conservation of energy is defined according to the following classical equation:

$$\frac{\partial \Omega_H}{\partial t} = - \nabla \cdot Q \tag{8.7}$$

where Ω_H, the heat content of the partly saturated soil per unit volume is defined as:

$$\Omega_H = H_c(T - T_r) + LnS_a\rho_v$$

T is temperature, T_r reference temperature, H_c heat capacity of the partly saturated soil, L the latent heat of vaporization, n the porosity, S_a the degree of saturation with respect to the air phase and ρ_v the density of the water vapour.

The heat capacity of unsaturated soil H_c at the reference temperature, T_r, can be expressed as:

$$H_c = (1 - n)C_{ps}\rho_s + n(C_{pl}S_l\rho_l + C_{pv}S_a\rho_v + C_{pda}S_a\rho_{da}) \tag{8.8}$$

where C_{ps}, C_{pl}, C_{pv} and C_{pda} are the specific heat capacities of solid particles, liquid, vapour and dry air respectively, ρ_s the density of the solid particles, ρ_{da} the density of the dry air, ρ_l the density of liquid water and S_l the degree of saturation with respect to liquid water.

The heat flux per unit area, Q, can be defined as:

$$\begin{aligned} Q = &- \lambda_T \nabla T + (v_v \rho_v + v_a \rho_v) L \\ &+ (C_{pl}v_l\rho_l + C_{pv}v_v\rho_l + C_{pv}v_a\rho_v + C_{pda}v_a\rho_{da})(T - T_r) \end{aligned} \tag{8.9}$$

where λ_T is the coefficient of thermal conductivity of the partly saturated soil and v_l, v_v and v_a are the velocities of liquid, vapour and air respectively.

The governing equation for heat transfer can be expressed, in primary variable form as:

$$\begin{aligned} C_{Tl}\frac{\partial u_l}{\partial t} &+ C_{TT}\frac{\partial T}{\partial t} + C_{Ta}\frac{\partial u_a}{\partial t} + C_{Tu}\frac{\partial \mathbf{u}}{\partial t} \\ &= \nabla[K_{Tl}\nabla u_l] + \nabla.[K_{TT}\nabla T] + \nabla[K_{Ta}\nabla u_a] + V_{Tl}\nabla u_l \\ &+ V_{TT}\nabla T + V_{Ta}\nabla u_a + V_{Tc}\nabla c_s + J_T \end{aligned} \tag{8.10}$$

where C_{Tj}, V_{Tj}, K_{Tj} and J_T are coefficients of the equation ($j = l, T, a, c_s, u$), u_l and u_a are the pore pressures for liquid and air respectively, and $\mathbf{u}$ is a nodal deformation vector.

Moisture transfer

The governing equation for moisture transfer in a partly saturated soil can be expressed as:

$$\frac{\partial(\rho_l n S_l)}{\partial t} + \frac{\partial(\rho_v n(S_l - 1))}{\partial t} = -\rho_l \nabla \cdot \mathbf{v}_l - \rho_l \nabla \cdot \mathbf{v}_v - \nabla \cdot \rho_v \mathbf{v}_a \qquad (8.11)$$

where t is the time. The velocities of pore liquid and pore air are calculated using a generalized Darcy's law, as follows:

$$\mathbf{v}_l = -K_l \left(\nabla \left(\frac{u_l}{\gamma_l} \right) + \nabla_z \right) + K_l^{c_s} \left(\nabla c_s \right) \qquad (8.12)$$

$$\mathbf{v}_a = -K_a \nabla u_a \qquad (8.13)$$

where K_l is the hydraulic conductivity, $K_l^{c_s}$ is the hydraulic conductivity with respect to chemical solute concentration gradient, K_a is the conductivity of the air phase, γ_l is the specific weight of water, u_l is the porewater pressure, z is the elevation, u_a is the pore air pressure and c_s the chemical solute concentration.

The inclusion of an osmotic flow term in the liquid velocity allows the representation of liquid flow behaviour found in some highly compacted clays. The definition of vapour velocity follows the approach presented by Philip-deVries. The governing equation for moisture transfer can be expressed, in primary variable form as:

$$C_{ll} \frac{\partial u_l}{\partial t} + C_{lT} \frac{\partial T}{\partial t} + C_{la} \frac{\partial u_a}{\partial t} + C_{lu} \frac{\partial u}{\partial t}$$
$$= \nabla [K_{ll} \nabla u_l] + \nabla [K_{lT} \nabla T] + \nabla [K_{la} \nabla u_a] + \nabla [K_{lc_s} \nabla c_s] + J_l \qquad (8.14)$$

where C_{lj}, K_{lj} and J_l are coefficients of the equation ($j = l$, T, a, c_s, u).

Dry air transfer

Air in partly saturated soil is considered to exist in two forms: bulk air and dissolved air. In this approach, the proportion of dry air contained in the pore liquid is defined using Henry's law.

$$\frac{\partial [S_a + H_s S_l] n \rho_{da}}{\partial t} = -\nabla \cdot [\rho_{da} (\mathbf{v}_a + H_s \mathbf{v}_l)] \qquad (8.15)$$

where H_s is the coefficient of solubility of dry air in the pore fluid.

The governing equation for dry air transfer can be expressed, in primary variable form, as:

$$C_{al} \frac{\partial u_l}{\partial t} + C_{aT} \frac{\partial T}{\partial t} + C_{aa} \frac{\partial u_a}{\partial t} + C_{au} \frac{\partial u}{\partial t}$$
$$= \nabla \cdot [K_{al} \nabla u_l] + \nabla \cdot [K_{aa} \nabla u_a] + \nabla \cdot [K_{ac_s} \nabla c_s] + J_a \qquad (8.16)$$

where C_{aj}, K_{aj} and J_a are coefficients of the equation ($j = l, T, a, c_s, u$).

Chemical solute transfer

When a chemical solute is considered non-reactive and sorption onto the soil surface is ignored, the governing equation for conservation of chemical solute can be defined as:

$$\frac{\partial (nS_l c_s)}{\partial t} = -\nabla \cdot [c_s v_l] + \nabla \cdot [D_h \nabla (nS_l c_s)] \qquad (8.17)$$

where D_h is the hydrodynamic dispersion coefficient defined as:

$$D_h = D + D_d \qquad (8.18)$$

D_h includes both molecular diffusion, D_d, and mechanical dispersion, D.

The governing equation for chemical solute transfer can be expressed, in primary variable form, as:

$$C_{c_s l} \frac{\partial u_l}{\partial t} + C_{c_s a} \frac{\partial u_a}{\partial t} + C_{c_s c_s} \frac{\partial c_s}{\partial t} + C_{c_s u} \frac{\partial u}{\partial t}$$
$$= \nabla \cdot [K_{c_s l} \nabla u_l] + \nabla \cdot [K_{c_s T} \nabla T] + \nabla \cdot [K_{c_s a} \nabla u_a]$$
$$+ \nabla \cdot [K_{c_s c_s} \nabla c_s] + J_{c_s} \qquad (8.19)$$

where $C_{c_s j}$, $C_{c_s j}$ and J_{c_s} are coefficients of the equation ($j = l, T, a, c_s, u$).

The approach has been extended to a multi-chemical species form with a sink term introduced to account for mass accumulation from reactions due to the sorption process. This is then coupled to a suitable geochemical model.

Stress–strain relationship

The total strain, ε, is assumed to consist of components due to suction, temperature, chemical and stress changes. This can be given in an incremental form, without loss of generality, as:

$$d\varepsilon = d\varepsilon_\sigma + d\varepsilon_{c_s} + d\varepsilon_s + d\varepsilon_T \qquad (8.20)$$

where the subscripts σ, c_s, T and s *refer* to net stress, chemical, temperature and suction contributions.

A number of constitutive relationships have been implemented to describe the contributions. In particular for the net stress, temperature and suction contributions, both elastic and elasto-plastic formulations have been employed. To describe the contribution of the chemical solute on the stress–strain behaviour of the soil, as a first approximation, an elastic state surface concept was proposed which described the contribution of the chemical solute via an elastic relationship based on osmotic potentials.

The governing equation for stress–strain behaviour can be expressed, in primary variable form as:

$$C_{ul}du_l + C_{uT}dT + C_{ua}du_a + C_{uc_s}dc_s + C_{uu}du + db = 0 \qquad (8.21)$$

where C_{uj} *are* coefficients of the equation ($j = l$, T, a, c_s, u) and b is the vector of body forces.

Numerical solution

A numerical solution of the governing differential equations is achieved by a combination of the finite element method for the spatial discretization and a finite difference time stepping scheme for temporal discretization. The Galerkin weighted residual method is employed to formulate the finite element discretization. For the flow equations a shape function N_m is used to define an approximation polynomial:

The general flow equation

$$\frac{\partial M}{\partial t} + \nabla \cdot q = 0 \qquad (8.22)$$

where M refers to mass or enthalpy and q refers to the flux term can be discretized via the Galerkin method as:

$$\int_\Omega \left(N_m^t \frac{\partial M}{\partial t} + \nabla N_m^t \cdot q \right) d\Omega + \int_{\Gamma_1} N_m^t q^* d\Gamma = 0 \qquad (8.23)$$

where q^* is a flux prescribed at boundary Γ_1.

The same method can be applied to the stress equilibrium equation. Applying this methodology to the governing differential equations, one obtains in a concise notation, the expression:

$$A\phi + B\frac{\partial \phi}{\partial t} + C = \{0\} \qquad (8.24)$$

where ϕ represents the variable vector $\{u_l,\ T,\ u_a\ c_s,\ u\}^t$.

To solve Eq. (8.24) a numerical technique is required to achieve temporal discretization. In this case an implicit mid-interval backward difference algorithm is implemented since it has been found to provide a stable solution for highly non-linear problems. This can be expressed as:

$$\mathbf{A}^{n+1/2}\boldsymbol{\phi}^{n+1} + \mathbf{B}^{n+1/2}\left[\frac{\boldsymbol{\phi}^{n+1} - \boldsymbol{\phi}^{n}}{\Delta t}\right] + \mathbf{C}^{n+1/2} = \{0\} \tag{8.25}$$

where the superscript represents the time level (the superscript n represents the last time interval and $n + 1$ represents the time level where the solution is being sought). Rearrangement of Eq. (8.25) yields:

$$\boldsymbol{\phi}^{n+1} = \left[\mathbf{A}^{n+1/2} + \frac{\mathbf{B}^{n+1/2}}{\Delta t}\right]^{-1}\left[\frac{\mathbf{B}^{n+1/2}\boldsymbol{\phi}^{n}}{\Delta t} - \mathbf{C}^{n+1/2}\right] \tag{8.26}$$

A solution can be found to Eq. (8.26) if **A**, **B** and **C** can be evaluated. This is achieved via an iterative scheme. For each iteration a revised set of **A**, **B** and **C** are calculated and the resulting solution checked for convergence against the following criteria:

$$\left|\frac{\boldsymbol{\phi}_{i}^{n+1} - \boldsymbol{\phi}_{(i-1)}^{n+1}}{\boldsymbol{\phi}_{(i-1)}^{n+1}}\right| < TL_{rel} \tag{8.27}$$

where TL_{rel} is the relative tolerance and i represents the iteration level. Successive iterations are performed until these criteria are satisfied. With appropriate initial and boundary conditions the set of non-linear coupled governing differential equations can be solved and the respective unknowns, that is, temperature, swelling pressure, porewater pressure and strain determined.

Note: One concludes that the code uses a number of the basic physical relationships valid for clay in bulk but the assumed transport and stress/strain mechanisms do not apply on the microstructural scale.

8.6.3 Code 'ABAQUS'

The finite element code ABAQUS [9] is used for calculating processes in a number of cases in SKB's R&D as has been discussed previously. While ABAQUS was originally designed for non-linear stress analyses, it has in its present form a capability for describing a large range of transient processes that are relevant to buffers and backfills. The code includes special material models for rock and soil, and can model geological formations with infinite boundaries and *in situ* stresses for example, by the medium's own weight. Detailed information of the available models, application of the code and the theoretical background are given in the ABAQUS Manuals.

The model used for water-unsaturated, swelling clay is described here to show similarities and differences from other codes used in the Prototype Repository Project.

Hydro-mechanical analysis

The hydro-mechanical model consists of a porous medium and wetting fluid and is based on equilibrium conditions, constitutive equations, energy balance and mass conservation using the effective stress theory. Equilibrium is expressed by writing the principle of virtual work for the volume under consideration in its current configuration at time t:

$$\int_V \sigma{:}\delta\varepsilon \, dV = \int_S \mathbf{t} \cdot \delta v \, dS + \int_V \hat{\mathbf{f}} \cdot \delta v \, dV, \tag{8.28}$$

where δv is a virtual velocity field, $\delta\varepsilon \overset{\text{def}}{=} sym(\partial\delta v/\partial \mathbf{x})$ is the virtual rate of deformation, σ is the true (Cauchy) stress, $\mathbf{t}$ are the surface tractions per unit area, and $\hat{\mathbf{f}}$ are body forces per unit volume. For our system, $\hat{\mathbf{f}}$ will often include the weight of the wetting liquid:

$$\mathbf{f}_w = S_r n \rho_w \mathbf{g}, \tag{8.29}$$

where S_r is the degree of saturation, n the porosity, ρ_w the density of the wetting liquid and $\mathbf{g}$ the gravitational acceleration, which we assume to be constant and in a constant direction (so that, for example, the formulation cannot be applied directly to a centrifuge experiment unless the model in the machine is small enough that g can be treated as constant). For the sake of simplicity we consider this loading explicitly so that any other gravitational term in $\hat{\mathbf{f}}$ is only associated with the weight of the dry porous medium. Thus, one can write the virtual work equation as:

$$\int_V \sigma{:}\delta\varepsilon dV = \int_S \mathbf{t} \cdot \delta v dS + \int_V \mathbf{f} \cdot \delta v \, dV + \int_V S_r n \rho_w \mathbf{g} \cdot \delta v \, dV, \tag{8.30}$$

where $\mathbf{f}$ are all body forces except the weight of the wetting liquid.

The simplified equation used in ABAQUS for the effective stress is:

$$\bar{\sigma}^* = \sigma + \chi u_w \mathbf{I} \tag{8.31}$$

where σ is the total stress, u_w the porewater pressure, χ a function of the degree of saturation (usual assumption $\chi = S_r$), and $\mathbf{I}$ the unitary matrix.

The conservation of energy implied by the first law of thermodynamics states that the time rate of change of kinetic energy and internal energy for a fixed body of material is equal to the sum of the rate of work done by the surface and body forces. This can be expressed as:

$$\frac{d}{dt}\int_V\left(\frac{1}{2}\rho\mathbf{v}\cdot\mathbf{v} + \rho U\right)dV = \int_s \mathbf{v}\cdot\mathbf{t}\,dS = \int_V \mathbf{f}\cdot\mathbf{v}\,dV, \tag{8.32}$$

where ρ is the current density, v is the velocity field vector, U is the internal energy per unit mass, t is the surface traction vector, f is the body force vector and V is the volume of solid material.

Constitutive equations

The constitutive equation for the solid is expressed as:

$$d\boldsymbol{\tau}^c = \mathbf{H}{:}d\boldsymbol{\varepsilon} + \mathbf{g} \tag{8.33}$$

where $d\boldsymbol{\tau}^c$ is the stress increment, $\mathbf{H}$ is the material stiffness, $d\boldsymbol{\varepsilon}$ is the strain increment and $\mathbf{g}$ is any strain independent contribution (e.g. thermal expansion). $\mathbf{H}$ and $\mathbf{g}$ are defined in terms of the current state, direction for straining, etc., and of the kinematic assumptions used to form the generalized strains.

The constitutive equation for the liquid (static) in the porous medium is expressed as:

$$\frac{\rho_w}{\rho_w^0} \approx 1 + \frac{u_w}{K_w} - \varepsilon_w^{th}, \tag{8.34}$$

where ρ_w is the density of the liquid, ρ_w^0 is its density in the reference configuration, $K_w(T)$ is the liquid's bulk modulus, and

$$\varepsilon_w^{th} = 3\alpha_w(T - T_w^0) - 3\alpha_w|_{T'}(T' - T_w^0) \tag{8.35}$$

is the volumetric expansion of the liquid caused by temperature change. Here $\alpha_w(T)$ is the liquid's thermal expansion coefficient, T is the current temperature, T' is the initial temperature at this point in the medium and T_w^0 is the reference temperature for the thermal expansion. Both u_w/K_w and ε_w^{th} are assumed to be small.

The mass continuity equation for the fluid combined with the divergence theorem implies validity of the pointwise equation:

$$\frac{1}{J}\frac{d}{dt}(J\rho_w S_r n) + \frac{\partial}{\partial \mathbf{x}}\cdot(\rho_w S_r n\mathbf{v}_w) = 0 \tag{8.36}$$

where J is the determinant of the Jacobian matrix and x is position.

The constitutive behaviour for pore fluid is governed by Darcy's law, which is generally applicable to low fluid velocities. Darcy's law states that, under uniform conditions, the volumetric flow rate of the wetting liquid through a unit area of the medium, $S_r n v_w$, is proportional to the negative of the gradient of the piezometric head:

$$S_r n v_w = -\hat{k}\frac{\partial\phi}{\partial x,} \tag{8.37}$$

where $\hat{k}$ is the permeability of the medium and ϕ is the piezometric head, defined as:

$$\phi \stackrel{\text{def}}{=} z + \frac{u_w}{g\rho_w} \tag{8.38}$$

where z is the elevation above some data and g the gravitational acceleration, which acts in the direction opposite to z. $\hat{k}$ can be anisotropic and is a function of the saturation and void ratio of the material. $\hat{k}$ is expressed in (m/s).

We assume that g is constant in magnitude and direction so:

$$\frac{\partial\phi}{\partial x} = \frac{1}{g\rho_w}\left(\frac{\partial u_w}{\partial x} - \rho_w g\right) \tag{8.39}$$

Vapour flow

Vapour flow is modelled as a diffusion process driven by a temperature gradient:

$$q_v = -D_{Tv}\frac{\partial T}{\partial x} \tag{8.40}$$

where q_v is the vapour flux and D_{Tv} the thermal vapour diffusivity.

Uncoupled heat transfer analysis

The basic energy balance is:

$$\int_v \rho\dot{U}\,dV = \int_s q\,dS + \int_v r\,dV \tag{8.41}$$

where V is volume of solid material with surface area S, ρ is the density of the material, $\dot{U}$ is the time rate of the internal energy, q is the heat flux per unit area of the body flowing into it and r is the heat supplied externally to the body per unit volume.

It is assumed that the thermal and mechanical processes are uncoupled in the sense that $U = U(T)$ only, where T is the temperature of the material, and q and r do not depend on the strains or displacements of the body, which makes the code different from COMPASS.

Constitutive equations

The relationship is usually written in terms of a specific heat, neglecting coupling between mechanical and thermal problems:

$$c(T) = \frac{dU}{dT}, \tag{8.42}$$

Heat conduction is assumed to be governed by the Fourier law:

$$\mathbf{f} = -\mathbf{k}\frac{\partial T}{\partial \mathbf{x}} \tag{8.43}$$

where $\mathbf{f}$ is the heat flux and $\mathbf{k}$ is the heat conductivity matrix, $\mathbf{k} = \mathbf{k}(T)$. The conductivity can be assumed to be anisotropic or isotropic.

Coupling of thermal and hydro-mechanical solutions

In ABAQUS, the coupled problem is solved through a 'staggered solution technique' as shown in Figure 8.11 and as described here:

- First a thermal analysis is performed where heat conductivity and specific heat are defined as functions of saturation and water content. In the first analysis these parameters are assumed to be constant and in the subsequent analyses they are read from an external file.
- The hydro-mechanical model calculates stresses, pore pressures, void ratios, degree of saturation etc. as functions of time. Saturation and void ratio histories are written onto an external file.
- The material parameters update module reads the file with saturation and void ratio data and creates a new file containing histories for saturation and water content. The saturation and water content histories are used by the thermal model in the subsequent analysis.

Steps 1–3 are repeated if parameter values are found to be different compared to those of the previous solution.

Description of the parameters of the material model of the buffer

Thermal flux that is modelled using the following parameters:

λ is the thermal conductivity.
c is the specific heat.

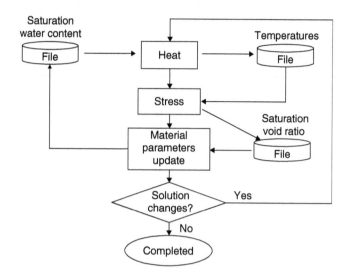

Figure 8.11 Heat transfer and hydro-mechanical calculations are decoupled. By using the iteration procedure schematically shown, the effects of a fully coupled THM model are achieved.

Flux of water in liquid form is modelled by applying Darcy's law with the water pressure difference as driving force in the same way as for water saturated clay. The magnitude of the hydraulic conductivity K_p of partly saturated clay is a function of the void ratio, the degree of saturation and the temperature. K_p is assumed to be a function of the hydraulic conductivity, K of saturated clay and the degree of saturation S_r according to Eq. (8.44):

$$K_p = (S_r)^\delta K \tag{8.44}$$

where K_p is the hydraulic conductivity of partly saturated soil (m/s), K is the hydraulic conductivity of completely saturated soil (m/s) and δ is the parameter (usually between 3 and 10).

Flux of water in vapour form is modelled as a diffusion process driven by the temperature gradient and the water vapour pressure gradient (at isothermal conditions) according to Eq. (8.45):

$$q_v = -D_{Tv}\nabla T - D_{pv}\nabla p_v \tag{8.45}$$

where q_v is the vapour flow, D_{Tv} is the thermal vapour flow diffusivity, T is the temperature, D_{pv} is the isothermal vapour flow diffusivity and p_v is the vapour pressure.

The thermal water vapour diffusivity D_{Tv} can be evaluated from moisture redistribution tests by calibration calculations. The following relations were found to yield acceptable results:

$$D_{Tv} = D_{Tvb} \quad 0.3 \leq S_r \leq 0.7 \tag{8.46}$$

$$D_{Tv} = D_{Tvb} \cdot \cos^a\left(\frac{S_r - 0.7}{0.3} \cdot \frac{\pi}{2}\right) \quad S_r \geq 0.7 \tag{8.47}$$

$$D_{Tv} = D_{Tvb} \cdot \sin^b\left(\frac{S_r}{0.3} \cdot \frac{\pi}{2}\right) \quad S_r \leq 0.3 \tag{8.48}$$

where a and b are factors that control the vapour flux at high and low degrees of saturation.

The diffusivity is thus constant with a basic value D_{Tvb} between 30% and 70% degree of saturation. It decreases strongly to $D_{Tv} = 0$ at 0% and 100% saturation. The influence of temperature and void ratio on the diffusivity is not known and not considered in the model. These assumptions have not been expressed or explained in microstructural terms.

Hydraulic coupling between the porewater and the pore gas

The pore pressure u_w of the unsaturated buffer material is always negative, in incompletely saturated material and is modelled in the ABAQUS case as a function of the degree of saturation S_r independently of the void ratio:

$$u_w = f(S_r) \tag{8.49}$$

This function is purely empirical and has not been derived or explained with reference to the transient microstructural constitution of buffer clay. ABAQUS allows for hysteresis effects, which means that two relationships can be derived (drying and wetting curves). They represent real conditions in bulk.

Mechanical behaviour of the particle skeleton

The mechanical behaviour is modelled using a non-linear Porous Elastic Model and a Drucker–Prager Plasticity model. The effective stress theory as defined by Bishop is applied and adapted to unsaturated conditions.

The *Porous Elastic Model* implies a logarithmic relation between the void ratio e and the average effective stress p according to Eq. (853).

$$\Delta e = \kappa \Delta \ln p \tag{8.50}$$

where κ is the porous bulk modulus and ν Poisson's ratio.

The *Drucker–Prager Plasticity* model contains the following parameters:

- β is the friction angle in the p–q plane,
- d is the cohesion in the p–q plane,
- ψ is the dilation angle and
- q is the $f(\varepsilon_{pl}^d) =$ yield function.

The yield function is the relation between Mises' stress q and the plastic deviatoric strain ε_p^d for a specified stress path. The dilation angle determines the volume change during shearing.

Thermal expansion

The volume change caused by the thermal expansion of water and particles can be modelled by use of the parameters:

- α_s is the coefficient of thermal expansion of solids (assumed to be 0).
- α_w is the coefficient of thermal expansion of water.

Only the expansion of the separate phases is taken into account. The possible change in volume of the combined structure by thermal expansion is not modelled. However, a thermal expansion of the porewater will change the degree of saturation, which in turn will change the volume of the whole structure.

Mechanical behaviour of the separate phases

The mechanical performance of the porewater and the particles are modelled as separate linearly elastic phases. The pore gas is not mechanically modelled.

Mechanical coupling between the microstructure and the pore water

The effective stress concept according to Bishop is used for modelling the mechanical behaviour of the water-unsaturated buffer material:

$$s_e = (s - u_a) + \chi(u_a - u_w) \tag{8.51}$$

with the following simplification:

u_a is the 0 (no account is taken to the pressure of enclosed gas)
χ is the S_r.

Required parameters

The required input parameters for the described THM model are the following:

- Thermal conductivity λ and specific heat c as function of void ratio e and degree of saturation S_r.
- Hydraulic conductivity of water saturated material K as function of void ratio e and temperature T.
- Influence of degree of saturation S_r on the hydraulic conductivity K_p.
- Basic water vapour flow diffusivity D_{vTb} and the parameters a and b.
- Matric suction u_w as a function of the degree of saturation S_r.
- Porous bulk modulus κ and Poisson's ratio ν.
- Drucker–Prager plasticity parameters β, d, ψ and the yield function.
- Bulk modulus (B_s) and coefficient of thermal expansion of water (B_w, α_w).
- Bishop's parameter χ (usual assumption $\chi = S_r$).
- Volume change correction ε_v as a function of the degree of saturation S_r (i.e. the 'moisture swelling' factor).

For the initial conditions one has the following parameters:

- Void ratio e
- Degree of saturation S_r
- Pore pressure u
- Average effective stress p.

Note: As for COMPASS, one concludes that the code uses a number of the basic physical relationships valid for clay in bulk but the assumed transport and stress/strain mechanisms do not apply on the microstructural scale. A major difficulty is that water transport in liquid form in unsaturated clay is assumed to take place by flow, neglecting the influence on molecular mobility of the void size and nearness to clay particle surfaces.

8.6.4 Code 'BRIGHT'

The CODE BRIGHT [10] is a finite element code for the analysis of THM problems in geological media developed by the Geomechanics group of the Geotechnical Engineering and Geosciences Department, Technical University of Catalunya – Centre for Numerical Methods in Engineering (UPC-CIMNE, Barcelona, Spain).

Basic approach

A porous medium composed of solid grains, water and gas is considered. Thermal, hydraulic and mechanical aspects are taken into account, including

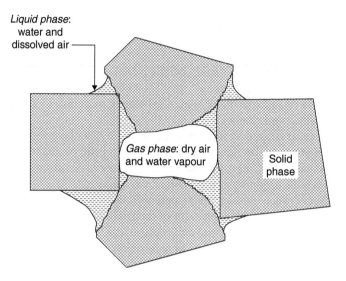

Figure 8.12 Schematic representation of an unsaturated porous material.

coupling between them in all possible directions. As illustrated in Figure 8.12 the problem is formulated in a multiphase and multispecies approach.

The three phases are:

- Solid phase (s)
- Liquid phase (l): water + air dissolved
- Gas phase (g): mixture of dry air and water vapour

The three species are:

- Solid (-):
- Water (w): as liquid or evaporated in the gas phase
- Air (a): dry air, as gas or dissolved in the liquid phase.

The following assumptions are considered in the formulation of the problem:

- Dry air is considered as a single species. It is the main component of the gaseous phase. Henry's law is used to express equilibrium of dissolved air.
- Thermal equilibrium between phases is assumed. This means that the three phases are at the same temperature.
- Vapour concentration is in equilibrium with the liquid phase, the psychrometric law expresses its concentration.

- State variables (also called unknowns) are: solid displacements, u (three spatial directions), liquid pressure P_l, gas pressure P_g and temperature, T.
- Balance of momentum for the medium as a whole is reduced to the equation of stress equilibrium together with a mechanical constitutive model to relate stresses with strains. Strains are defined in terms of displacements.
- Small strains and small strain rates are assumed for solid deformation. Advective terms due to solid displacement are neglected after the formulation is transformed in terms of material derivatives (in fact, material derivatives are approximated as Eulerian time derivatives). In this way, volumetric strain is properly considered.
- Balance of momentum for dissolved species and for fluid phases are reduced to constitutive equations (Fick's law and Darcy's law).
- Physical parameters in constitutive laws are function of pressure and temperature. For example: concentration of vapour under planar surface (in psychrometric law), surface tension (in retention curve), dynamic viscosity (in Darcy's law), are strongly dependent on temperature.

The governing equations that CODE-BRIGHT solves are:

- Mass balance of solid
- Mass balance of water
- Mass balance of air
- Momentum balance for the medium
- Internal energy balance for the medium.

Associated with this formulation is a set of necessary constitutive and equilibrium laws. Table 8.1 is a summary of the constitutive laws and equilibrium restrictions that have been incorporated in the general formulation. The dependent variables that are computed using each law are also included.

The constitutive equations establish the link between the independent variables (or unknowns) and the dependent variables. There are several categories of dependent variables depending on the complexity with which they are related to the unknowns. The governing equations are finally written in terms of the unknowns (indicated in Table 8.2) when the constitutive equations are substituted in the balance equations.

The resulting system of partial differential equations is solved numerically. The numerical approach is divided into two parts: spatial and temporal discretization. The finite element method is used for the spatial discretization while finite differences are used for the temporal discretization. The latter is linear in time and the implicit scheme uses two intermediate points, $t^{k+\varepsilon}$ and $t^{k+\theta}$ between the initial t^k and final t^{k+1} times. Finally, since the problems are non-linear, the Newton–Raphson method has been adopted to find an iterative scheme.

Table 8.1 Constitutive equations and equilibrium restrictions

Equation	Variable name
Constitutive equations	
Darcy's law	Liquid and gas advective flux
Fick's law	Vapour and air non-advective fluxes
Fourier's law	Conductive heat flux
Retention curve	Liquid phase degree of saturation
Mechanical constitutive model	Stress tensor
Phase density	Liquid density
Gases law	Gas density
Equilibrium restrictions	
Henry's law	Air dissolved mass fraction
Psychrometric law	Vapour mass fraction

Table 8.2 Equations and summary of variables

Equation	Variable name	Variable
Equilibrium of stresses	Displacements	u
Balance of water mass	Liquid pressure	P_l
Balance of air mass	Gas pressure	P_g
Balance of internal energy	Temperature	T

Processes modelled

The processes modelled are:

- T and TM modelling
- HM and THM modelling of rock mass
- THM modelling of buffer, backfill and interaction with near-field rock
- C modelling of buffer, backfill and groundwater.

The mechanical stress–strain relationship of the bentonite is defined by means of an elastoplastic model specially designed for unsaturated soil behaviour [11]. The code deals with chemical processes like complexation, oxidation/reduction reactions, acid/base reactions, precipitation/dissolution of minerals, cation exchange, sorption and radioactive decay. The total analytical concentrations are adopted as basic transport variables and chemical equilibrium is achieved by minimizing Gibbs Free Energy.

Note: As for COMPASS and ABAQUS, one concludes that the code uses a number of the basic physical relationships valid for clay in bulk but the assumed transport and stress/strain mechanisms do not apply on the microstructural scale. A major difficulty is that water transport in liquid form in unsaturated clay is assumed to take place by flow, neglecting the influence on molecular mobility of the void size and nearness to clay particle surfaces.

8.6.5 Code 'THAMES'

Code THAMES [11], developed in Japan, can be used for describing processes in a saturated–unsaturated porous medium. The code has been validated by use of data from laboratory tests, engineered scale tests and *in situ* experiments. The code formulations are described here in some detail since they appear to be in fair agreement with actual transport properties on the microstructural level.

Processes modelled

The finite element code THAMES is used for analysis of coupled thermal, hydraulic and mechanical processes in a saturated–unsaturated porous medium. The unknown variables are: (a) total pressure, (b) displacement vector and (c) temperature. The quadratic shape function is used for the displacements and a linear one for total pressure and temperature.

Major governing equations in THM-modelling

The mathematical formulation for the model utilizes Biot's theory, with the Duhamel–Neuman's form of Hooke's law, and an energy balance equation. The governing equations are derived with fully coupled thermal, hydraulic and mechanical processes.

Assumptions

The governing equations are derived under the following assumptions:

- The medium is poro-elastic.
- Darcy's law is valid for the flow of water through a saturated–unsaturated medium.
- Heat flow occurs only in solid and liquid phases. The phase change of water from liquid to vapour is not considered.
- Heat transfer among three phases (solid, liquid and gas) is disregarded.
- Fourier's law holds for heat flux.
- Water density varies depending upon temperature and the pressure of water.

Governing equations

The standard equation of motion is written in a total stress expression as:

$$\sigma_{ij,j} + \rho b_i = 0 \tag{8.52}$$

where σ_{ij} is the stress, ρ the density of a soil-water system and b_i the body force.

The Bishop–Blight extension of the Terzaghi effective stress relationship to include saturated–unsaturated media is used in the following form:

$$\sigma_{ij} = \sigma'_{ij} + \chi\delta_{ij}\rho_f g\psi \tag{8.53}$$

where σ'_{ij} is the effective stress, δ_{ij} is the Kronecker delta, ρ_f is the unit weight of water, g is the acceleration of gravity and ψ the pressure head. Subscript f means 'fluid'. Parameter χ is defined as:

$$\chi = 1 \quad \text{(Saturated zone)}, \quad \chi = \chi(S_r) \quad \text{(Unsaturated zone)} \tag{8.54}$$

χ is a non-linear function of S_r (the degree of saturation). As for ABAQUS, it is approximated as S_r.

The following general equilibrium equation for the effective stress is obtained:

$$(\sigma'_{ij} + \chi\delta_{ij}\rho_f g\psi)_{,j} + \rho b_i = 0 \tag{8.55}$$

where $(\chi\delta_{ij}\rho_f\, g\psi)$ is a term implying that changes in the pressure head influence the equilibrium equation.

The effects of temperature can be implemented in a constitutive law by using Duhamel–Neuman's relationship for solid media. For an isotropic, linear elastic material the following constitutive law is obtained:

$$\sigma'_{ij} = C_{ijkl}\varepsilon_{kl} - \beta\delta_{ij}(T - T_o) \tag{8.56}$$

where $\beta = (3\lambda + 2\mu)\alpha_T$. C_{ijkl} is the elastic matrix, ε_{kl} the strain tensor, T the temperature, λ and μ Lamé's constants and α_T the thermal expansion coefficient. Subscript o means that the parameter is in a reference state.

The infinitesimal strain-deformation relationship is:

$$\varepsilon_{kl} = \tfrac{1}{2}(u_{k,l} + u_{l,k}) \tag{8.57}$$

where u_i is the deformation vector.

The stress equilibrium equation is obtained that takes into account the effects of temperature and pore pressure changes:

$$\left[\tfrac{1}{2}C_{ijkl}(u_{k,l} + u_{l,k}) - \beta\delta_{ij}(T - T_o) + \chi\delta_{ij}\rho_f g\psi\right]_{,j} + \rho b_i = 0 \tag{8.58}$$

where $(-\beta\delta_{ij}(T - T_o))_j$ is a term which stands for the influence of heat transfer on the equilibrium equation.

Although further expressions have been derived to take into account the compressibility of the porewater, these are omitted herein because of limitations of space.

Energy conservation law

Considering the existence of an unsaturated zone in the buffer, the equation of energy conservation is written as:

$$nS_r\rho_f C_{vf}\left(\frac{\partial T_f}{\partial t} + V_f\nabla T_f\right) = -\nabla nS_r J_f - \left(\frac{\partial P}{\partial T_f}\right)_{\rho_f} nS_r T_f\nabla V_f \qquad (8.59)$$

where C_v is the specific heat and J is the heat flux by conduction. The first term on the left-hand side shows the time dependency of energy, the second term the change in energy due to heat convection. The first term on the right-hand side expresses the change in energy by heat conduction and the second term the reversible energy change caused by compression.

It is assumed that the movement of water through the buffer clay in the saturation phase is so slow and the surface areas of all phases so large that local thermal equilibrium among the various phases is achieved instantaneously. This means that the heat transfer between phases in the ground can be disregarded. Assuming that Fourier's law is valid for heat conduction, the following equations apply:

$$\begin{aligned} J_f &= -K_{Tf}\nabla T \\ J_s &= -K_{Ts}\nabla T \end{aligned} \qquad (8.60)$$

where K_T is the coefficient of heat conduction.

Equation (8.61) is the energy conservation law in which the effects of stress-deformation and ground water flow are considered.

$$K_{Tm} = nS_r K_{Tf} + (1 - n)K_{TS} \qquad (8.61)$$

The first, second and third terms on the right-hand side express changes in energy due to heat conduction, pore water pressure and reversible energy caused by solid deformation, respectively.

Governing equations

The following expressions represent the governing equations for the coupled thermal, hydraulic and mechanical problem:

$$\left[\tfrac{1}{2}C_{ijkl}(u_{k,l} + u_{l,k}) - \beta\delta_{ij}(T - T_o) + \chi\delta_{ij}\rho_f g\psi\right]_{,j} + \overline{\rho}_s b_i = 0$$

$$\{\rho_f k(\theta)_{ij} h_{,j}\}_{,i} - \rho_{fo} n S_r \rho_f g \beta_P \frac{\partial h}{\partial t} - \rho_f \frac{\partial \theta}{\partial \psi} \frac{\partial h}{\partial t} - \rho_f S_r \frac{\partial u_{i,i}}{\partial t}$$

$$+ \rho_{fo} n S_r \beta_T \frac{\partial T}{\partial t} = 0 \tag{8.62}$$

$$(\rho C_v)_m \frac{\partial T}{\partial t} + n S_r \rho_f C_{vf} V_{fi} T_{,i} - K_{Tm} T_{,ii}$$

$$- n S_r T \frac{\beta_T}{\beta_P} k(\theta) h_{,ii} + \frac{1}{2}(1-n)\beta T \frac{\partial}{\partial t}(u_{i,j} + u_{j,i}) = 0 \tag{8.63}$$

where $\bar{\rho}_s = (1-n)(\rho_s - S_r\rho_f)$ and ρ_s is the density of a solid phase.

To calculate the water/vapour movement and heat-induced water movement in the buffer, the continuity equation used in THAMES is as follows:

$$\left\{ \xi\rho_1 D_\theta \frac{\partial\theta}{\partial\psi} (h_{,i} = z_{,i}) + (1-\xi)\frac{\rho_1^2 g K}{\mu_1} h_{,i} \right\} + \{\rho_1 D_T T_{,i}\}_{,i}$$

$$- \rho_{1_o} n S_r \rho_1 g \beta_P \frac{\partial h}{\partial t} - \rho_1 \frac{\partial\theta}{\partial\psi} \frac{\partial h}{\partial t} - \rho_1 S_r \frac{\partial u_{i,i}}{\partial t} + \rho_{1_o} n S_r \beta_T \frac{\partial T}{\partial t} = 0 \tag{8.64}$$

where D_θ is the isothermal water diffusivity, θ is the volumetric water content, ψ is the water potential head and K is the intrinsic permeability. The symbol ξ is the unsaturated parameter with $\xi = 0$ in the saturated zone and $\xi = 1$ in the unsaturated zone. The symbol μ_1 is the viscosity of water, ρ_1 the density of water and g the gravitational acceleration. D_T is the thermal water diffusivity, n the porosity, S_r the degree of saturation, β_P the compressibility of water, β_T the thermal expansion coefficient of water and z the elevation head. u_i is the displacement vector, T temperature, h the total head and t time. The subscript o means the reference state. The equation means that the water flow in the unsaturated zone is expressed by the diffusion equation and in the saturated zone by the Darcy's law. These definitions are in good agreement with modern physical water migration theories (cf. Chapter 6).

The equilibrium equation takes the swelling behaviour into account as defined by Eq. (8.65):

$$\left[\frac{1}{2}C_{ijkl}(u_{k,l} + u_{l,k}) - F\pi\delta_{ij} - \beta\delta_{ij}(T - T_o) + \chi\delta_{ij}\rho_1 g(h - z) \right] + \rho b_i = 0 \tag{8.65}$$

where C_{ijkl} is the elastic matrix, ρ the density of the medium and b_i the body force. χ is the parameter for the effective stress, $\chi = 0$ in the unsaturated zone with $\chi = 1$ in the saturated zone. The symbol F is the coefficient related to the swelling pressure and $\beta = (3\lambda + 2\mu)\alpha_s$, where λ and μ are Lame's constants and α_s the thermal expansion coefficient.

The swelling pressure π is a function of water potential head (ψ) as follows:

$$\pi(\theta_1) = \rho_l g(\Delta\psi) = \rho_l g(\psi(\theta_1) - \psi(\theta_0)) = \rho_l g \int_{\theta_0}^{\theta_1} \frac{\partial\psi}{\partial\theta}d\theta \qquad (8.66)$$

where θ_0 is the volumetric water content in the initial state. This is compatible with the assumption that swelling pressure is equivalent to the water potential.

Numerical techniques

The Galerkin type finite element technique is employed to formulate a finite element discretization. In order to obtain stable solutions, linear isoparametric elements are used to represent the behaviour of total head h and temperature T. Quadratic isoparametric elements are used to express displacement u_i. In order to integrate time derivatives, a time weighting factor is introduced and thus, any type of finite difference scheme may be applied.

Note: In contrast to the other codes, the assumed transport of water and vapour according to THAMES does apply on the microstructural scale.

8.6.6   Code 'Rockflow/Rockmech' (RF/RM)

The following principles are basic to the model used by BGR, Germany [12]:

- Effective stress and consolidation concepts according to Biot and Terzaghi.
- Mohr–Coulomb failure concept including the influence of internal friction, cohesion and dilatancy is valid.
- Drucker–Prager's model of the first invariant of the total stress tensors and the second invariant of the stress deviators is valid.
- Roscoe/Schofield/Burland's Clay Model is applied.

The model realistically uses completely coupled thermo/hydraulic/mechanical processes considering the non-linear effects caused by that is, permeation under unsaturated conditions and elasto-plastic behaviour of the buffer clay. For the buffer, the influence of desiccation fractures, swelling and microstructural changes is also taken into account. Major processes in the saturation and subsequent percolation of the buffer are:

- Reduction of the permeable pore space by the expansion of the smectite clay particles and thereby an associated drop in bulk hydraulic conductivity.
- Changes in effective stress and strain in the saturation phase affect the mechanical behaviour of the clay. Here, changes in temperature, water

content and stress conditions in both the buffer clay and confining rock play a major role.
- Formation and transport of vapour.
- Osmosis.

The most important hydrothermal effects caused by thermal and pressure gradients are related to:

- redistribution of the initial porewater content in the buffer including vapour formation and condensation;
- changes in viscosity and hydraulic conductivity of water in different temperature regions;
- influence on porewater pressure and saturation rate by the groundwater pressure in the surrounding rock;
- change of the heat conductivity of the buffer in the saturation phase;
- chemical alteration of the porewater and mineral phases is realized but not considered in the model.

Basic equations

- Flow of two fluid phases (compressible and incompressible fluids) in a deformable thermo-poro-elastic porous medium based on Biot's consolidation concept is considered. No phase change of water beyond 100°C is considered.
- There are four unknown field functions to determine: gas pressure p_g, water pressure p_W, solid displacements u and equilibrium temperature T. They are all defined and applied in the calculations.

8.7 Applicability of models

8.7.1 *Preconditions for application of the numerical models*

General

The conditions for applying the models described in the preceding section depend very much on the initial and boundary conditions, the heat source and the changes with time that it may undergo and whether the material properties and characteristics change because of water uptake and chemical changes in the system. In practical cases the most obvious controlling condition is the access to water for saturation of the initially partly saturated clay. We will examine the applicability of the models by referring to the KBS-3V case and will use the clay buffer data and hydrological characteristics that have been selected for modelling the integrated performance of the near-field system [13].

Uptake of water from the rock – hydraulic boundaries

Uptake of water from the rock and backfill is the THMC process that ultimately yields complete water saturation and the final density distribution. If the host rock permits very little water to be transported to the deposition holes, the backfill may serve as the major water source. Again, it is necessary to point out the great significance of the rate of water saturation for the physical and chemical maturation of the buffer.

Water transport from the backfill to the buffer is controlled by the water pressure and by the suction of the buffer, as well as by the degree of water saturation of the backfill. The water pressure in the permeable backfill at the contact between buffer and backfill may rise relatively quickly if it has access to water from intersected hydraulically active fracture zones. If not, the rise in water pressure will be relatively low and slow.

The rock around the deposition holes gives off water through discrete hydraulically and mechanically active or activated discontinuities. The hydraulic transport capacity depends on the frequency and conductive properties of the discontinuities, while the distribution over the periphery of the holes is controlled both by the location of the intersecting fractures and by the conductivity of the shallow boring-disturbed zone (EDZ).

Hydration of the buffer and backfill by uptake of water from the rock is associated with expansion of the upper part of the buffer and consolidation of the backfill as well as with axial displacement of the canisters.

Wetting has a C-effect since Ca^{2+}, Mg^{2+}, SO_4^{-2} and Cl^- will be precipitated in the form of minerals with reversed solubility such as sulphates and carbonates at the wetting front. These will move towards the canister and will inevitably interact with the smectite minerals and the canister. The *temperature gradient* has a C-effect since silicate minerals including the smectite component will dissolve. The extent of the dissolution is greater at locations near the heater, and less at regions close to the rock. This differential will result in the migration of the released silica towards the host rock boundary, and accumulation of precipitates in the outer part of the bentonite. Silicification causes cementation and brittleness of the clay. The major consequences of the water uptake are:

- Water uptake provides the desired condition of the canisters of complete embedment by a practically impermeable clay medium. The sooner complete saturation is obtained the sooner drops the high buffer temperature near the hot canister and hence stops the enrichment of salt near it.
- The canister may undergo displacement in the course of the wetting of the buffer and backfill. The rate of their respective expansion can be quite different and cause significant canister movements.
- Expansion of the upper part of the buffer and compression of the contacting backfill will change their hydraulic conductivities. Thermally or tectonically induced compression and shearing of hydraulically

active rock discontinuities may have an impact on the ability of the rock to transmit water to the buffer in the deposition holes.

• If the rate of water uptake is slow, salt precipitation in the buffer close to the canisters is enhanced. This also enhances corrosion of the copper canisters. Exposure to vapour given off from the wetting front will affect the clay minerals and cause dissolution and precipitation of silica.

*The piezometric and temperature conditions
at the AEspoe site*

Pressure and temperature data obtained before the start of the buffer-and-backfill applications form the basis for the various THMC modelling attempts. Data made available in the course of the project (prototype experiment) was available for use for calibration and updating of the models.

The average bulk hydraulic conductivity of the near-field rock around the deposition holes is K = E–12 to K = 4E–10 m/s. The average inflow of water in the holes is less than 0.006 l/min in all the holes except Hole 1, which has an inflow of 0.08 l/min.

The water pressure at about 2 m distance from the tunnel wall is between 100 kPa and 1.5 MPa. Varying the hydraulic conductivity of the backfill from E–11 to E–9 to m/s in calculations gave corresponding water pressure distributions in the rock. Figure 8.13 shows the distribution of pressure and salinity on the level where the field test is conducted.

8.8   Application of the codes to the AEspoe case

8.8.1   General

The THMC models discussed in Section 8.5 have been used to predict the evolution of the EBS. This section provides the most recent results obtained and presented by the respective modelling groups. In the interest of clarity and simplicity, the complete data set was intended to concern only the 'wettest' hole (No. 1). However, this was extended to also cover Hole 3 by two of the modellers. We will confine ourselves here to the presentation of results from four of the models – leaving ABAQUS out since the premises respecting the heat source did not agree with the actual figures. All data and results are derived from the two EC-supported projects 'Prototype Repository Project' and 'CROP' (cf. [14, 15]).

8.8.2   Predictions for comparison with
actual recordings

The geometry of the model for 3D FEM calculations is shown in Figure 8.14. Special definitions respecting number of considered holes and boundary conditions are specified by the respective modelling group.

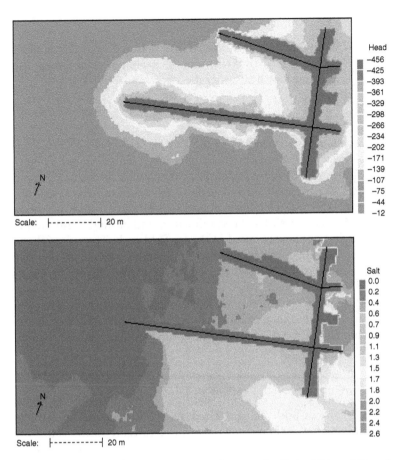

Scale: |----------| 20 m

Head
-456
-425
-393
-361
-329
-298
-266
-234
-202
-171
-139
-107
-75
-44
-12

Salt
0.0
0.2
0.4
0.6
0.7
0.9
1.1
1.3
1.5
1.7
1.8
2.0
2.2
2.4
2.6

Scale: |----------| 20 m

Figure 8.13 Pressure head (m) in the upper figure and salinity (%) at 447 m depth [I. Rhen]. The pressure and salinity are lowest in the near-field of the tunnels.

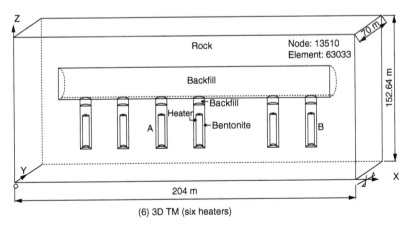

(6) 3D TM (six heaters)

Figure 8.14 Schematic view of generalized 3D model (THAMES).

The calculations have been based on material data obtained from laboratory tests provided by SKB, that is, an initial rock temperature of 12–14°C, and a constant heat generation per canister of 1800 W.

Compass

A full three-dimensional model of the Prototype Repository incorporating all of the primary features of the tunnel has been developed. The model domain measures 200 m by 100 m by 200 m and is shown in Figure 8.15a. This model has been discretized using 8 noded hexahedral elements and consists of 158 175 elements and 146 380 nodes and is shown in Figure 8.15b. The mesh has been refined in and around the buffer with a coarser mesh discretization used in the far-field rock. The size of the model has been reduced by 50% via the introduction of a vertical symmetry plane along the centre of the tunnel and hence the computational requirements of the model are considerably reduced. This geometrical model has been used for the majority of the numerical modelling work with smaller three-dimensional and two-dimensional models being implemented to investigate the mechanical response of the buffer and pellets under thermal and hydraulic gradients.

Temperature at mid-height canister

Figure 8.16 shows both the simulated and experimentally measured temperature plots after 900 days for the 3 different radii positions in Borehole 1 at the mid-height of the canister. It can be seen that there is excellent agreement between the sets of results. At a radius of 0.585 m, the temperature has been simulated to be 71.5°C after 717 days. The corresponding experimentally measured value at the same position is 71.4°C. At a radius of 0.685 m, the temperature is simulated as 64.8°C after 717 days, with a

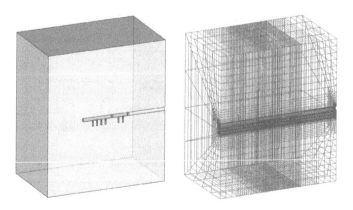

Figure 8.15 (a) 3D tunnel domain and (b) 3D tunnel mesh.

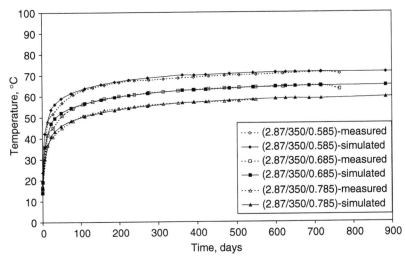

Figure 8.16 Measured and simulated temperature plots for mid-height canister in Hole 1. Note that the radial positions given as the last three digits are in metres, for example, 0.585 m for the first set of (top) points. The lower curve represents the conditions at the rock.

corresponding experimentally measured value of 65°C. Finally, at a radius of 0.785 m, there were no results measured after approximately 550 days when the temperature was measured to be 58.2°C. The simulated value at the same position and similar time was 58.3°C. These results illustrate that the temperature regime is well understood and captured by the model for the 'wet' Hole 1.

Hydration of buffer at mid-height canister

The hydraulic conductivity of the granite was taken as 10^{-11} (E–11) m/s with a representative fracture intersecting the borehole with a hydraulic conductivity of 10^{-9} (E–9) m/s. Figure 8.17 shows the simulated and experimentally measured degree of saturation plots for three different radii positions at the mid-height of the canister. It can be seen that initially there is an over prediction of drying in the buffer. At a radius of 0.785 m, a minimum value of 78.6% is reached after approximately 20 days. Experimentally the buffer closest to the rock exhibits immediate recharge following heater activation with the pellet-filled region appearing to have very little effect in terms of retarding the rate of resaturation in the more centrally located buffer. After 100 days the correlation between the experimental and simulated results is much improved, with almost complete saturation being achieved after approximately 400 days in both cases.

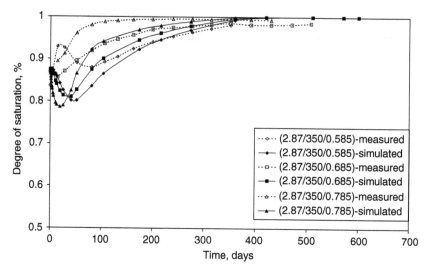

Figure 8.17 Measured and simulated degree of saturation for mid-height canister in Hole 1. Radial positions are given as the last three digits in the legend (upper curve representing the conditions at the rock).

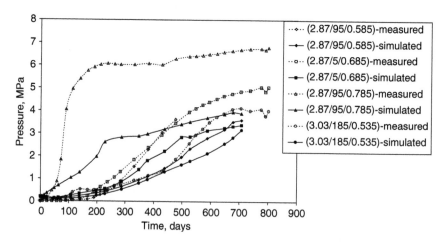

Figure 8.18 Measured and simulated total pressure plots for mid-height canister in Hole 1. Radial positions distant from the canister are given as the last three digits in the legend (upper curve represents the conditions close to the rock).

Total pressure in buffer

Figure 8.18 shows both the simulated and experimentally measured total pressure plots in the buffer at four different positions in Hole 1.

There is good qualitative correlation between the results. Quantitatively, the simulation has obtained close agreement with measured swelling pressures in the buffer – except in the region close to the rock/buffer interface where the buffer experiences the highest swelling pressure by recharge from the granite. Closest to the canister surface the swelling pressure was simulated as reaching 3.2 MPa compared to, a measured pressure of 4.1 MPa after 700 days. At the location closest to the rock/buffer interface the simulated pressure of 3.9 MPa can be compared with the experimentally measured value of 6.7 MPa. It is believed that this difference is due to an overestimation of the compressibility of the pellet region leading to a relief of some of the swelling pressure developed on saturation.

Whereas Hole 1 can be taken as 'wet' all other deposition holes in the Prototype Repository test drift have very little inflow of water and probably represent the majority of deposition holes in a repository constructed in crystalline rock. The hydration rate is of particular interest and it was found that for the recorded average flow of water to the 'dry' Hole 3 drying takes place in the buffer as a consequence of the movement of vapour away from the hotter regions closest to the heater. At $r = 0.585$ m the predicted degree of saturation would reach a minimum of 71.8% after 225 days, which follows the measured trend.

This behaviour can also be seen in Figure 8.19 at $r = 0.685$ m for both the simulated and measured results but with less overall drying taking place due to the larger distance to the heater. At $r = 0.785$ m there is an initial over-prediction of drying that may be a result of an over-estimation of suction in the nearby pellet region. An overall conclusion is that both

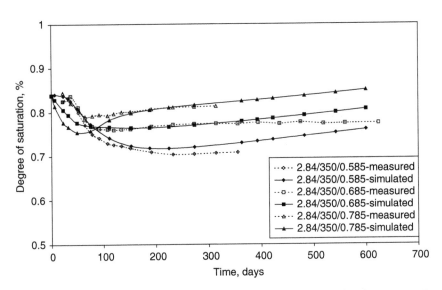

Figure 8.19 Measured and simulated saturation plots for mid-height canister in Hole 3 (the driest one). The upper curve after 200 days represents the conditions close to the rock. The fit between measured and calculated data is fairly good.

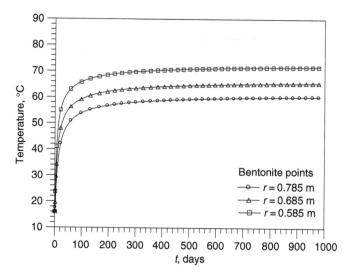

Figure 8.20 Evolution of temperature computed for three points in the buffer in the 'wet' Hole 1: close to the heater (upper curve), close to the rock (lower curve) and mid-point.

predicted and actual wetting rates indicate that complete water saturation will require many tens of years to centuries.

Bright

Temperature at mid-height canister   Figure 8.20 summarizes the numerical results, including the evolution of temperature for three typical points in the buffer.

   Hydration of buffer at mid-height canister   Figure 8.21 presents the evolution of degree of saturation against time for the mid plane of Hole 1. According to the calculations saturation would be practically complete after 1000 days. This is in accord with the outcome of the *COMPASS* simulations and with the actual observed buffer evolution rate.

   Total pressure in buffer   The computed pressure evolution results accorded with the COMPASS simulations and with actual recorded values.

Thames

Temperature at mid-height canister   Figure 8.22 shows the predicted temperature evolution at the selected points. We observe that the temperature steadily increases with time and does not reach steady state conditions even after 1000 days.

   Hydration of buffer at mid-height canister   The initial water content of 17% for the buffer represents a degree of saturation of approximately 60%,

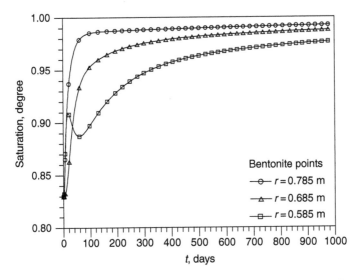

Figure 8.21 Predicted degree of saturation at mid-height of heater number 1 (the wettest one). Upper curve corresponds to a point close to the rock, and the lower one to a point close to the heater. Near the rock saturation is practically complete (99%) after 1000 days.

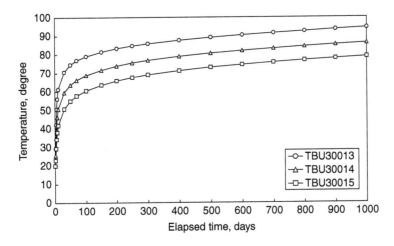

Figure 8.22 Predicted temperature at mid-height canister in Hole 1 (highest close to the rock).

and a corresponding relative humidity of about 80%. The evolution of the saturation is shown in Figure 8.23. In contrast to the other models no drying is expected at the canister.

Total pressure in buffer Figure 8.24 shows the change in total stress, or in practice, the swelling pressure in the buffer at mid-height canister in Hole 1.

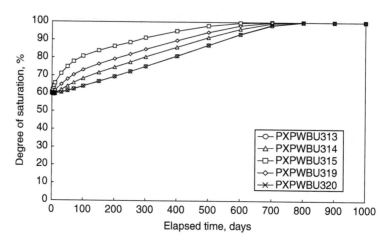

Figure 8.23 Predicted degree of saturation of the buffer at mid-height canister in Hole 1 (Highest close to the rock).

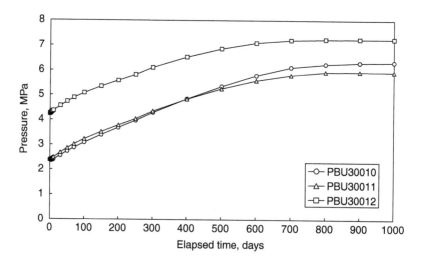

Figure 8.24 Predicted total stress at mid-height canister in Hole 1. Curve –o– represents buffer contacting the canister.

RF/RM

Hydration and swelling pressure Figure 8.25 shows the change in degree of water saturation at mid-height of a deposition hole intersected by 2 assumed vertical water-bearing fractures and without such fractures respectively. The ambient water pressure was assumed to be 1 MPa, driving water into the fracture and further to the excavation-disturbed zone (EDZ). One finds that complete water saturation requires many decades in the fracture case and at least a hundred years for the fracture-free case. Still, for

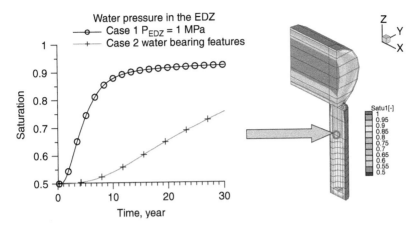

Figure 8.25 Water saturation of buffer and backfill versus time for two relevant cases (RF/RM BGR). The central part of the tunnel backfill and the buffer still has low degrees of water saturation in the picture.

the first-mentioned case a high degree of saturation is reached already after about 8–10 years. The pressure at the canister surface reaches and proceeds beyond 5 MPa after 3 years.

8.8.3 *Actual recordings*

The actual *temperature* evolution is shown in Figure 8.26. One finds that it reached about 72°C in the clay adjacent to the canister surface and around 60°C at the rock after about 700 days, or around 2 years. The average temperature gradient is about 0.34 centigrade per centimetre radial distance.

The actual *hydration* process is shown in Figure 8.27. It indicates that the clay had reached RH values of 92–94%, which represents approximately the degree of water saturation after about one year.

The actual *total pressure* changed as shown by Figure 8.28 indicating that complete maturation will require more than 2.5 years, that is, longer than indicated by the RH measurements. The latter may indicate too rapid saturation caused by water leakage along the cables to the RH metres.

8.8.4 *Comparison of predictions and recordings*

Temperature

The predicted and actual temperatures at the rock and the canister surface at mid-height canister level are shown in Table 8.3. One finds that two of the models give adequate data whereas one (THAMES) somewhat exaggerates the temperature. For RF/RM, the canister temperature was selected and hence controlled the heat evolution of the entire buffer.

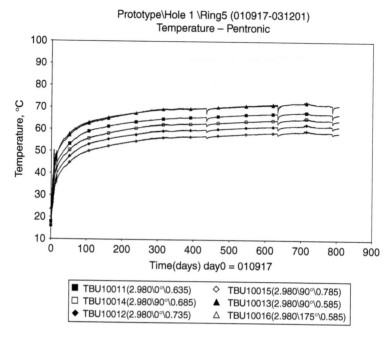

Figure 8.26 Temperature at mid-height of the canister in Hole 1 (highest *T* close to the rock).

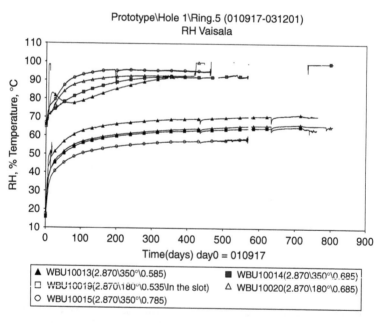

Figure 8.27 RH and temperature distributions in the buffer at mid-height canister in Hole 1 (highest RH and lowest *T* close to the rock). The upper curve set shows the RH readings.

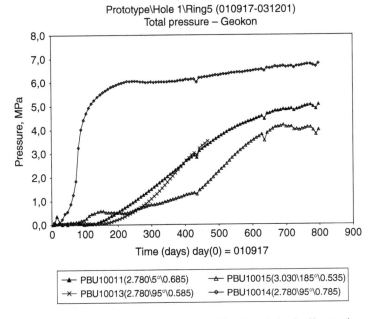

Figure 8.28 Evolution of total pressure at mid-height of the buffer in the wettest hole. The highest pressure, about 6.7 MPa, was reached after about 2 years close to the rock while the lowest (4 MPa) was obtained from a cell close to the canister.

Table 8.3 Actual and expected temperature in centigrade at the canister and rock surfaces at mid-height in the wettest hole after 1–2 years from start

Location	Recorded	COMPASS (UWC)	BRIGHT (CIMNE, ENRESA)	RF/RM (BGR)	THAMES (JNC)
Canister	1 y = 69	1 y = 70	1 y = 70	1 y = 100[a]	1 y = 87
	2 y = 72	2 y = 72	2 y = 72	2 y = 100[a]	2 y = 92
Rock	1 y = 56	1 y = 56	1 y = 57	1 y = 92[a]	1 y = 71
	2 y = 60	2 y = 59	2 y = 60	2 y = 96[a]	2 y = 76

Notes
a Set by modeller.
b Resulting from the fixed 100°C temperature.

Hydration

The predicted and actual degrees of saturation agree fairly well for all the models except RF/RM and THAMES as shown by Table 8.4. However, the latter gave best fit of all models for the canister after two years. It should be noted that all the models gave quicker saturation than in reality.

Table 8.4 Actual and expected degree of water saturation at the canister and rock surfaces at mid-height in the wettest hole after 1–2 years from start

Location	Recorded	COMPASS (UWC)	BRIGHT (CIMNE, ENRESA)	RF/RM (BGR)	THAMES (JNC)
Canister	1 y = 92	1 y = 96	1 y = 95	1 y = 76	1 y = 79
	2 y = 92	2 y = 100	2 y = 97	2 y = 84	2 y = 99
Rock	1 y = 92	1 y = 98	1 y = 99	1 y = 95	1 y = 94
	2 y = 92	2 y = 100	2 y = 99	2 y = 98	2 y = 100

Table 8.5 Actual and expected pressure in MPa at the canister and rock surfaces at mid-height in the wettest hole after 1–2 years from start

Location	Recorded	COMPASS (UWC)	BRIGHT (CIMNE, ENRESA)	RF/RM (BGR)	THAMES (JNC)
Canister	1 y = 1.0	1 y = 0.8	1 y = 3.0	1 y = 3.5	1 y = 4.7
	2 y = 4.0	2 y = 3.2	2 y = 5.1	2 y = 4.8	2 y = 6.2
Rock	1 y = 6.0	1 y = 2.8	1 y = 5.0	1 y = 3.5	1 y = 6.4
	2 y = 6.7	2 y = 3.9	2 y = 7.2	2 y = 4.8	2 y = 7.2

Total pressure

Table 8.5 shows the predicted total pressure which deviates significantly from the actual data in some cases. Thus, for COMPASS the predicted pressure growth was too slow.

8.8.5   Conclusions from the comparison of predictions and recordings

General

- All the theoretical models provided calculated results that are on the same order of magnitude as the measurements and can be used for rough prediction of the temperature, hydration and pressure build-up in buffer of the type used in the Prototype Repository Project.
- Good agreement between predictions and measurements is obtained for the temperature evolution. Some models overestimate the temperature for the first two years, hence yielding a safe, conservative prediction, whereas the others give very accurate forecasting. There are indications that the thermal conductivity of the buffer is higher than assumed and that the heat transfer is assisted by some undefined mechanism like convection through vapour flow.
- Almost all the models provide calculated results that are fairly well in agreement with the recordings. They provide information on the wetting

rate for deposition holes with 'unlimited' access to water from the rock that can be useful for scoping purposes. The predicted rate of saturation is generally too high. This indicates that all processes involved in water uptake are not fully understood and therefore not been properly reflected in the models.

- For deposition holes with limited access to water for hydration the situation is more uncertain. One of the models provided fairly accurate predictions of hydration in a 'dry' hole by basing its calculations on measured inflow before placement of the buffer in the hole (No. 3). The comparison of calculated and observed results indicates that it is more difficult to predict the wetting process for a repository with much less access to water.

- The evolution of pressure and mechanical response of the buffer is the most difficult task because it requires that fracturing and displacements in the buffer are included in the models and that the interrelation of hydration/dehydration and swelling/drying are relevant. Since prediction of the hydration rate appears to be uncertain, forecasting of the mechanical response is even more uncertain. However, the models manage to provide calculated results that can be usefully used as scoping data. As in the case of hydration, some pressure gauges may have reacted too soon because of water migration along cables and this may imply that the maturation of the buffer is in fact even slower than indicated by the recordings.

A note on instrumentation

A first and major problem in recording accurate hydration data in experiments such as the Prototype Repository Project, and any field test concerning the evolution of top and bottom liners of waste landfills, is the risk of water migration along cables to moisture sensors. This type of activity may have given incorrect information on the rate of hydration of the uninstrumented buffer. Accordingly, the reference values to be compared with the predictions may not be adequate. This is a very serious matter and must be carefully addressed when field performance is to be modelled and predicted by both experimentalists and modellers.

A note on the microstructural impact

In principle, the calculated results from the models tested are in agreement with the conceptual models for the macroscopic behaviour of the buffer insofar as heat and mass transfer are concerned. This however does not allow one to conclude that the microstructural controls on macroscopic properties and functions have been, or are correspondingly well-accounted for. When a significant number of empirically obtained material data are obtained on a macroscopic basis, and if these used to adjust the models to fit measured or expected data, one should expect some accord between calculated and measured system performance. This procedure does not fit the Class A prediction rules traditionally invoked for predictions of performance of target systems or facilities. For the Prototype Repository project, it is apparent that only blind calculations based on fundamental thermodynamic relationships can provide one with the required level of confidence in the models and the manner in which they are applied.

Of particular importance is the assumed and modelled hydration of the initially incompletely saturated buffer clay. The models, except THAMES, are based on the assumption that hydration takes place by flow under a hydraulic gradient caused by the suction of the still unsaturated part of the buffer. The THAMES code, on the other hand, assumes flow in the saturated part and diffusive water uptake driven by the gradient in water content, which is more in line with the basic hydration/dehydration mechanisms described earlier in the book.

8.9   Models for special purposes

8.9.1   *Ion diffusion*

Porewater chemistry

The fact that migration of water molecules and hydrated or unhydrated chemical elements dissolved species is of fundamental importance for a number of microstructurally related processes, such as hydration, cation exchange and chemical reaction rates in clay seals. We consider here recent models for describing and predicting ionic transport in clays like liners and buffers [15, 16].

The starting point is the porewater chemistry, which, for any soil, results from complex interactions between the solution and the minerals. Equilibrium modelling can explain the basic chemical reactions such as dissolution, precipitation and ion exchange.

Ion migration and sorption in porous medium
with narrow channels

The small space between mineral surfaces in dense smectite clay means that the ion distribution and mobility adjacent to these surfaces play a major role in modelling ion diffusion. MGC (modified Gouy–Chapman) theory and the conventional GC theory can be used for calculating the negative potential and the interaction of electrical double-layers but it is clear from analyses of dielectric permittivity profiles that the potential can only be defined for the mid-point between clay platelets. The major flaw of the GC theory is that it implies unreasonably high cation concentrations close to the mineral surfaces. The MGC is therefore a preferable scientific tool [16].

The basic expression for ion diffusion flux J has the following form:

$$J = nD_p(dc/dx) \tag{8.67}$$

where n is the porosity, D_p is the diffusion constant in water and dc/dx is the concentration gradient.

Since smectites have a sorption potential that affects the diffusion rate, the diffusion flux one uses is the so-called apparent diffusivity D_a:

$$D_a = nD_p/(n + K_d\rho_d) \tag{8.68}$$

where K_d is the distribution factor and ρ_d is the dry density

For the specific purpose of modelling ion diffusion through narrow channels one can express the ion diffusion flux J by the expression in Eq. (8.78) [10]:

$$J = \phi\tau^{-2}D[\eta^{-1}(1 + \gamma)]\Gamma \tag{8.69}$$

where ϕ is the surface potential, τ is the tortuosity factor, D is the diffusion coefficient, η is the measure of viscosity of surface-near shear zone and Γ = surface-excess charge at distance γ from the surface.

In this model K_d appears in the following form:

$$K_d = \phi\Gamma/(1 - \phi)\rho_d \tag{8.70}$$

Defined in this way K_d accounts for negative adsorption, that is, repulsion and exclusion of ions [16]. Based on the microstructural modelling described in Chapter 4, it is assumed the clay microstructure consists of an array of parallel capillaries with periodic stepwise changes in diameter as in Figure 8.29 and with dimensions according to Table 8.6. The figure gives the relative amounts of monovalent anion concentration in percent in a REV for different concentrations. It illustrates that Donnan exclusion is valid not only for the interlamellar space but also between aggregates and stacks of lamellae if the distance between mineral surfaces is sufficiently small.

Ion flux capacity on the microstructural scale

Naturally, the stochastic nature of the microstructure implies that both the width and straightness of the channels vary and that the density of the gel fillings is not a constant. The channel size and gel density vary over any individual cross section in isotropic clay, still yielding approximately the same microstructural F_2-value for each of them. This parameter is hence a useful quantitative measure of the conductivity-controlling microstructural component. Since it also represents the part of the clay in which pore

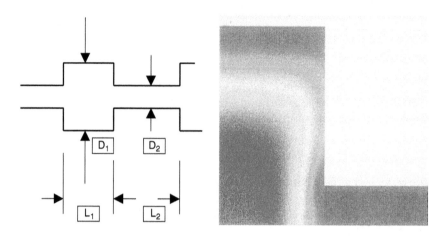

Figure 8.29 Relative monovalent anion concentration (%) in a REV for equilibrating 0.1 M monovalent electrolyte in MX-80 clay with the density 2130 kg/m³. The picture shows that anions (dark area in the big void) are confined in the centre of the wide part of the widest channels (50 μm) and that only cations (gray area adjacent to void boundaries and in the narrow void) are located close to the mineral surfaces and in the channel [17].

Table 8.6 Channel dimensions

Bulk density at water saturation, kg/m³	D_1, Å	D_2, Å	L_1, Å	L_2, Å
2130	50	10	100	50
1850	100	20	200	100
1570	500	30	1000	500

diffusion takes place it is a determinant of the diffusion capacity as well. Thus, the ratio of void volume available for uniaxial diffusion, and total volume should be approximately proportional to F_2 for anions and to unity for non-reactive cations like sodium.

The ratio of the ion diffusion capacity, for example, iodine and sodium should hence be about 0.17 for a bulk density of 2000 kg/m³, which is about 4 times higher than the experimentally determined ratio. The main reason for this discrepancy is that channel tortuosity and constrictions have not been accounted for. Hence, when applying the correction of F_2 used in calculating the hydraulic conductivity caused by constrictions one arrives at a ratio for iodine and sodium diffusion capacity that is very close to the recorded value.

8.9.2 *Geochemical transients during and after buffer maturation*

General

The matter of mineral changes in soil exposed to changed porewater chemistry and temperature is classical and the literature provides an immense amount of calculations based on commercially available geochemical simulation codes such as PHREEQE. The complexity of many geochemical processes requires development of codes for solving problems that deal with reactive chemical transport and a code of this sort, ARASE, has been developed by Grindrod and Takase [15]. It will be described here in a condensed form, referring to these authors.

Background

In the course of the wetting of the buffer in the deposition holes, there is a source of both oxygen and carbonate species due to dissolution of minerals. At the saturation, both aerobic and anaerobic corrosion of the buffer/canister interface is possible, the transport taking place both through advection and diffusion.

The ARASE code makes it possible to consider a number of aqueous and mineral species that are relevant to metal corrosion and mineral changes [15]. Also, dissolution/precipitation causing changes in rheological behaviour and coupled changes in water and gas transport capacities can be calculated.

Model equation

Processes of primary interest occur in the water saturated part of the buffer, which is taken to be radially symmetric for repository concepts like the KBS-3V. Aqueous species are assumed to be advected by groundwater flow and by diffusion.

The starting point is to specify a system of reaction-diffusion-advection equations for n solute species, with concentrations in mols per unit fluid volume denoted by $u = (u_1, \ldots, u_n)^T$ and m solid species with concentrations $w = (w_1, \ldots, w_n)^T$, where T refers to transport. Both u and w distributions depend on spatial location, denoted x, and time, t. The porosity is termed $\theta = \theta(x, t)$. v will denote the Darcy flow rate (v/θ is the mean fluid velocity).

Distinction is made between fast, instantaneous and slow reactions, which are treated kinetically. p fast reactions $[F(u) = (F_1, \ldots, F_p)^T]$ and q slow reactions $[S(u,w) = (S_1, \ldots, S_q)^T]$ are assumed to take place. Hence, if the j_{th} fast reaction is of the form in Eq. (8.81):

$$\alpha[u_1] + \beta[u_2] \quad \Rightarrow \quad \gamma[u_3] + \delta[u_4] \tag{8.71}$$

where α, β, γ and δ are stoichiometric constants, the reaction proceeds to the right or to the left according to whether the term F_j in Eq. (8.82) is positive or negative:

$$F_j = a_1^\alpha a_2^\beta K + a_3^\gamma a_4^\delta \tag{8.72}$$

Here, $a_i = a_1(u_i)$ is the activity of species i.

The rate of reaction is given by $R_f F_f$ where R_f is a large positive constant. For fast reactions $F = 0$. If the jth slow reaction involves precipitation/ dissolution of the form in Eq. (8.82) then the corresponding slow kinetic reaction term is that in Eq. (8.73):

$$\alpha u_1 + \beta u_2 \Rightarrow \gamma w_q \tag{8.73}$$
$$S_j = k_{1j}\alpha_1^\alpha \, \alpha_2^\beta + k_{2j}A(W_1) \tag{8.74}$$

where the k-factors are the rates of forward and back reactions and A the activity of the mineral species w_1, which depends upon specific area (S_j) among other features.

Mass conservation implies that:

$$(\eta u) = T(u) + \eta B_s S(w, u) + \eta B_F R F(u) \tag{8.75}$$

where $T(u)$ denotes the net rate of increase in mass per unit volume due to transport of solutes (diffusion and advection). η is the porosity of the porous medium, $R = \text{diag}(R_1, \ldots, R_p)$ is a $(p \times p)$ matrix of fast reaction rates, and B_F and B_S are $(n \times p)$ and $(n \times q)$ matrices containing the stoichiometric coefficients of the fast and slow reactions, respectively.

For complete reactions with $t = \infty$, the RF terms must be eliminated which yields a new system governing the $n - p$ degrees of freedom. One introduces an $n \times (n - p)$ matrix C such that $C^T B_F = 0$. This yields, after some rearrangements, the relationship in Eq. (8.76):

$$(\eta U)_t = C^T T(u) + \eta C^T B_S S(w, u) \tag{8.76}$$

At any time $t > 0$ one solves Eq. (8.86) together with the algebraic system in Eqs (8.77) and (8.78):

$$0 = F(u) \tag{8.77}$$
$$U = C^T u \tag{8.78}$$

which yields u in terms of U. In general Eq. (8.78) must be coupled with the equations for the mineral species:

$$w_t = B_S(w, u) \quad (w > 0) \tag{8.79}$$

where B_s denotes the stoichiometry matrix for the mineral precipitation and dissolution reactions.

The mineralogy distribution may be coupled back to the transport terms $T(u)$ via the volumetric change and the displacement of pore water. One has:

$$\eta = g(w) \tag{8.80}$$

where g is some known function describing porosity in terms of the current mineralogy, and porewater mass conservation implies:

$$(\eta_t + \nabla\Delta)\cdot v = 0 \tag{8.81}$$

where v satisfies Darcy's law:

$$v = - k(\eta)\Delta\cdot P \tag{8.82}$$

where k is the permeability, which depends on the porosity and may be altered by the chemical processes, and P is the water pressure.

In these terms one can write:

$$T_t = \Delta(h\eta D_t\Delta u_j - u_i) \tag{8.83}$$

where D_t is the diffusion–dispersion tensor for the ith species.

The approach is to solve Eqs (8.79)–(8.83) simultaneously by employing a non-linear solver with variable time stepping. The result is the change in mineral composition accompanied by precipitation and cementation or increase in porosity and hydraulic conductivity. As to the boundary conditions one must impose conditions on the solute species, that is, flux or concentration conditions.

Example of impact of boundary conditions on the illitization of smectite buffer

Applying codes like ARASE, one is able to predict chemical reactions such as illitization of smectite buffer clay as a function of the access to potassium. An example of the outcome of such calculations, applying the geotechnical code FLAC, gives one the time-dependence of the illitization as shown by Figure 8.30. The basis was the earlier described Pytte-type model that is currently used in SKB's predictions of long-term conversion of smectite to illite, that is, the hypothesis that transformation takes place either via mixed-layer S/I formation or as neoformation of illite. As explained in Section 7.3.2, the selection of a relevant activation energy figure is a crucial thing, another one being the definition of the hydraulic and hence also the chemical boundaries of the near-field. Thus, as in the Kinnekulle case, access to potassium in the groundwater can be the rate-controlling factor.

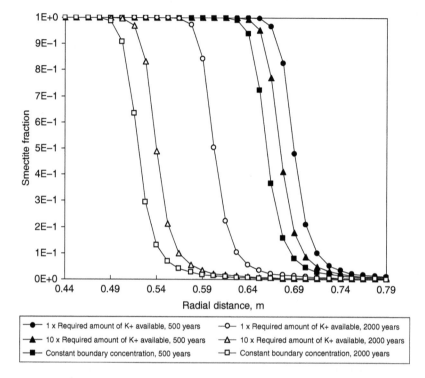

Figure 8.30 Effects of limitations in potassium availability to the illitization of buffer smectite in KBS-3V. Smectite fraction profiles obtained as a function of time with the initial smectite content being E+0. (Harald Hökmark, Clay Technology AB, Sweden.)

The example illustrated in Figure 8.30 concerned the KBS-3V case with the inner, hot boundary of the buffer clay being 0.44 m off the center of the deposition hole and the outer colder boundary being 0.79 m away from the centre. The illustration shows that for the required potassium content to yield illitization, about 50% will be converted in the middle of the buffer after 2000 years. The calculation was based on a high activation energy for S-to-I alteration and gave, for some of the boundary conditions, a far more conservative rate of conversion to the non-expandable mineral than the earlier showed calculations based on Grindrod/Takase's model. Still, it serves to illustrate the importance of the assumed access to potassium, the key parameter for the S-to-I degradation of normally heated smectite buffers.

8.10 References

1. Smith, W.O., 1943. 'Thermal transfer of moisture in soils'. *American Geophysical Union Transactions*, Vol. 2, No. 5: 11–523.

2. Gurr, C.G., Marshall, T.J. and Hutton, J.T., 1952. 'Movement of water in soil due to a temperature gradient'. *Soil Science*, Vol. 74: 335–345.

3. Rollins, R.L., Spangler, M.G. and Kirkham, D., 1954. 'Movement of soil moisture under a thermal gradient'. In the *Proceedings of Highway Research Board*, Vol. 33: 492–508, Washington.

4. Kuzmak, J.M. and Serada, P.J., 1957. 'The mechanism by which water moves through a porous material subjected to a temperature gradient: 2. Salt tracer and streaming potential to detect flow in the liquid phase'. *Soil Science*, Vol. 84: 419–422.

5. Yong, R.N. 2001. *Geoenvironmental Engineering: Contaminated Soils, Pollutant Fate, and Mitigation*, CRC Press, Boca Raton, FL, 307p.

6. Sposito, G., 1984. *The Surface Chemistry of Soils*, Oxford University Press, New York, 223p.

7. Yong, R.N. and Xu, D.-M., 1988. 'An identification technique for evaluation of phenomenological coefficients in coupled flow in unsaturated soils'. *International Journal for Numerical and Analytical Methods in Geomechanics*, Vol. 12: 283–299.

8. Thomas, H.R. and King, S.D., 1991. 'Coupled temperature/capillary potential variations in unsaturated soil'. *ASCE, Journal of Engineering Mechanics* Vol. 117, No. 11: 2475–2491.

9. Börgesson L. 2001. Äspö Hard Rock Laboratory: Selection of THMCB models. In SKB International Progress Report IPR-01-66. Swedish Nuclear Fuel and Waste Management Co (SKB) Stockholm. pp. 46–54.

10. Ledesma, A., and Chen, G., 2003. 'T-H-M modelling of the Prototype Repository experiment at Äspö, Hard Rock Laboratory', Sweden. In the *Proceedings of Geoproc Conference*, KTH, Stockholm (Sweden), pp. 370–375.

11. Chijimatsu, M., and Sugita Y., 2001. Äspö Hard Rock Laboratory: Selection of THMCB Models. In SKB International Progress Report IPR-01-66. Swedish Nuclear Fuel and Waste Management Co (SKB). SKB, Stockholm. pp. 37–46.

12. Liedtke, L., 2002. *The Saturation and Resulting Swelling Pressure of a Deposition Hole – Tunnel System Filled with Bentonite under Consideration of the Excavation Disturbed Zone with the help of two phase flow theory – ANDRA; Clays in Natural and Engineered Barriers for Radioactive Waste Confinement*. Rheims.

13. Svemar, C., 2005. Prototype Repository Project. Final Report of European Commission Contract FIKW-2000-00055, Brussels, Belgium.

14. Svemar, C., 2000 Cluster Repository Project (CROP) Final Report of European Commission Contract FIR1-CT-2000-20023, Brussels, Belgium.

15. Grindrod, P. and Takase H., 1993. 'Reactive chemical transport within engineered barriers'. In the *Proceeding of the Fourth International Conference on the Chemistry and Migration Behaviour of Actinides and Fission Products in the Geosphere*, Charleston, SC USA, 12–17 December. R.Oldenburg Verlag 1994. pp. 773–779.

16. Pusch, R., Muurinen, A., Lehikoinen, J., Bors, J. and Eriksen, T., 1999. Microstructural and Chemical Parameters of Bentonite as Determinants of Waste Isolation Efficiency. Final Report of European Commission Contract F14W-CT95-0012, Brussels, Belgium.

Index

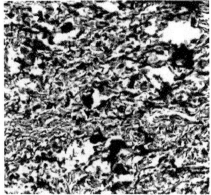

Colour Plate I Micrograph of thin section of acrylate-embedded mixture of 10% MX-80 bentonite and 90% crushed TBM muck. The picture shows coatings (black) of the rock fragments by a thin layer of very small quartz particles, which strongly increase the permeability of the mixture. Magnification 100× [5] (see Figure 4.21).

Colour Plate II Example of digitalized micrograph of Wyoming bentonite (MX-80) with a bulk density at saturation of 1800 kg/m^3 (dry density 1270 kg/m^3). Black = densest parts of clay matrix *a*. Grey represents component *b* and white represent open parts of this component. Edge length is 3 μm (see Figure 3.31).

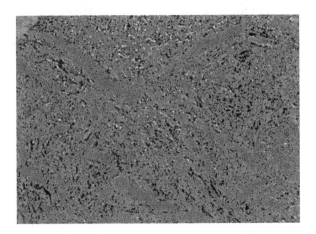

Colour Plate III Digitalized TEM micrograph of HDP$_y^+$-treated MX-80 clay with 1760 kg/m^3 density in water saturated form. The width of the micrograph is 1 mm. Wide voids forming channels with up to 100 μm width separate the dense and homogeneous blocks of aggregates [5] (see Figure 4.16).

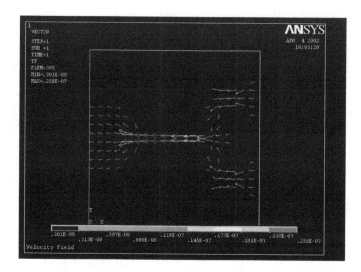

Colour Plate IV FEM calculation in 2D of flow through 30 × 30 μm² clay element of smectitic clay with 1600 kg/m³ density. For a hydraulic gradient of 100 the maximum flow rate represented by the largest vectors is 2E–6 m/s [5] (see Figure 5.9).

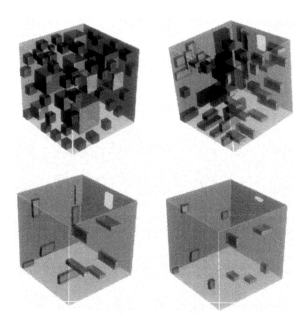

Colour Plate V 3D system of boxes representing voids that are open or filled with soft clay gels in a cubical clay element with 30 μm edge length. Upper left: Bulk density of 1300 kg/m³ with a high frequency of 1–20 μm voids, maximum void size 50 μm. Upper right: Bulk density of 1600 kg/m³ with a moderately high frequency of 1–15 μm voids, maximum size 20 μm. Lower left: Bulk density of 1800 kg/m³ with a low frequency of 1–10 μm, maximum size 15 μm. Lower right: Bulk density of 2000 kg/m³ with very few voids, all smaller than 5 μm (see Figure 5.24).

Colour Plate VI View of open pit mining of Friedland Ton (FIM GmbH). The pit is about 100 m deep (see Figure 2.23).

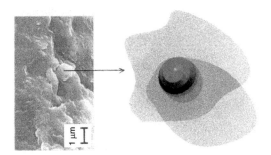

Colour Plate VII Bacterium embedded in montmorillonite clay with a dry density of 1500 kg/m³. Left: SEM picture, right: schematic view of the protruding bacterium, magnified to some extent (see Figure 5.23).

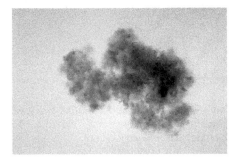

Colour Plate VIII Micrograph of 20 μm aggregate of illite/mixed-layer particles in low-electrolyte river water (Elbe) photographed by optical microscopy (Photo: Kasbohm) (see Figure 3.13).

Colour Plate IX Growth of soft clay through the perforation of an 80 mm copper tube with dense bentonite in an oedometer – to simulate maturation of a borehole plug. After 8 hours, the central part of the bentonite core is still unaffected by water. After a few days, the soft gel is densified by clay moving from the core through the perforations (see Figure 4.13).

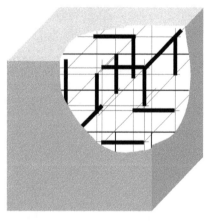

Colour Plate X Shaly Canadian bentonite exploited for manufacturing clay powder. The clay is cemented but crushing in the processing plant breaks down the strongest bonds (Photo: M. Gray) (see Figure 5.25).

Colour Plate XI Schematic view of the 3D conceptual model (see Figure 5.12).

Colour Plate VI View of open pit mining of Friedland Ton (FIM GmbH). The pit is about 100 m deep (see Figure 2.23).

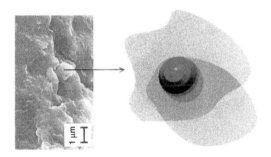

Colour Plate VII Bacterium embedded in montmorillonite clay with a dry density of 1500 kg/m^3. Left: SEM picture, right: schematic view of the protruding bacterium, magnified to some extent (see Figure 5.23).

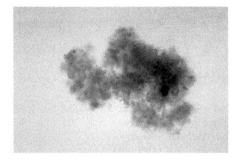

Colour Plate VIII Micrograph of 20 μm aggregate of illite/mixed-layer particles in low-electrolyte river water (Elbe) photographed by optical microscopy (Photo: Kasbohm) (see Figure 3.13).

Colour Plate IX Growth of soft clay through the perforation of an 80 mm copper tube with dense bentonite in an oedometer – to simulate maturation of a borehole plug. After 8 hours, the central part of the bentonite core is still unaffected by water. After a few days, the soft gel is densified by clay moving from the core through the perforations (see Figure 4.13).

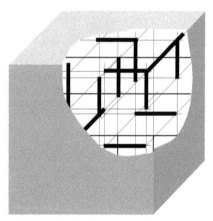

Colour Plate X Shaly Canadian bentonite exploited for manufacturing clay powder. The clay is cemented but crushing in the processing plant breaks down the strongest bonds (Photo: M. Gray) (see Figure 5.25).

Colour Plate XI Schematic view of the 3D conceptual model (see Figure 5.12).

Milton Keynes UK
Ingram Content Group UK Ltd.
UKHW021631071024
449327UK00020BA/1276